0, 2, 7, 23, 200·

AIR MONITORING SURVEY DESIGN

AIR MONITORING SURVEY DESIGN

by

KENNETH E. NOLL

Professor of Environmental Engineering
Illinois Institute of Technology
Chicago, Illinois

and

TERRY L. MILLER

President
Enviro-Measure
Knoxville, Tennessee

ANN ARBOR SCIENCE
PUBLISHERS INC
P.O. BOX 1425 • ANN ARBOR, MICH. 48106

PREFACE

This book is intended to provide the information necessary to allow comprehensive air monitoring surveys to be designed in a systematic and cost-effective manner. It is generally understood that air quality monitoring involves equipment selection, calibration and operation. It is less well understood, however, that a complex air quality survey also involves (a) the selection of monitoring sites, (b) the integration of air monitoring and meteorological equipment into a compatible network, and (c) a comprehensive quality assurance program involving instrument calibration, accurate data recording and comprehensive data evaluation.

First, the book provides the overview and definitions which are necessary to an understanding of Air Quality Monitoring Survey design. Step-by-step provisions for conducting air monitoring studies are provided in Chapter III. Then, it describes specific site selection methods for point, line, and area air monitoring stations bound upon general solutions to diffusion models. Further, the book provides specific information to allow the selection of the appropriate air sample collecting and analysis methods, as well as detailed procedures for instrument calibration. Chapter XI provides price ranges and commercial sources for air monitoring hardware.

Another major area covers step-by-step procedures for conducting meteorological surveys. The availability and use of historical data as well as methods of collection and evaluation of meteorological data are identified in detail.

The final chapters provide methods for general environmental data evaluation, comparison of air quality data to Air Quality Standards, validation of mathematical simulation models and statistically determining the number of samples needed to accurately define the mean and maximum pollution concentration expected at an air monitoring station.

K. E. Noll
Chicago, Illinois

v

DEDICATED IN MEMORY OF

Chester A. Gordon
and
Rubin Silver

CONTENTS

CHAPTER I

INTRODUCTION

Air quality is a dynamic and complex environmental phenomenon exhibiting large temporal and spatial variation. The temporal and spatial variations in atmospheric levels of pollution, which is the essence of air quality, are caused by (a) changes in the rate at which sources emit material and (b) changes in meteorological and topographic conditions which contribute to the dilution of the material, provide for chemical reactions in the atmosphere, and control the removal of the various pollutants.

Monitoring surveys designed to characterize the air quality of an area can become complex because they are required to provide data to allow a resolution of the dynamic nature of air quality in terms of temporal and spatial variation. Collecting and analyzing a sufficient number of representative samples of ambient air to allow an evaluation of air quality requires the application of advanced technology in the areas of analytical chemistry; operation and calibration of electrochemical and meteorological instruments; and data recording, collecting and processing. Extensive meteorological and source strength information is required if meaningful data interpretation and valid prediction of future air quality is to be provided. Furthermore, air quality concentrations will be relatively high near sources but will become diluted and undergo transformation with distance. Thus, a wide range of values require measurement, and analytical methods must be sensitive enough to measure highly diluted material.

In recent years, ambient air monitoring has become an important part of many environmental impact statements to satisfy the requirements of the National Environmental Policy Act.[1] This has made air quality surveys even more complex, requiring adequate planning to assure that prescribed objectives can be attained in the shortest possible time and at the least cost. Comprehensive survey design and management is also

1

required because monitoring is expensive and time consuming, requiring skilled personnel and sophisticated analytical equipment.

THE ROLE OF MONITORING AND MODELING IN AIR QUALITY ASSESSMENT

In order to determine the environmental impact of a new or existing project, provide routine source surveillance or operate an intermittent air pollution control program, it is necessary to estimate air pollution concentrations either by monitoring or modeling or by a combination of these methods. The interrelationship between monitoring and modeling, which can lead to an optimum design of an air quality survey, is illustrated schematically in Figure 1.

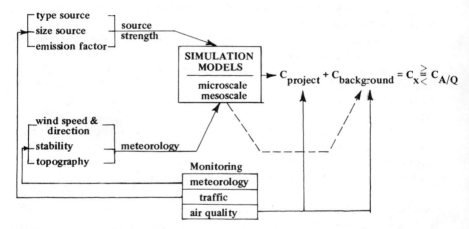

Figure 1. Interaction of modeling and monitoring to assess highway air quality impact.

As shown in the figure, the purpose of mathematical models is to quantitatively combine the effects of source strength and meteorology to describe the resulting ambient air pollution concentrations. Source strength is affected by a number of variables including the size of the source, variable emission rates, and the efficiency of air pollution control equipment employed. Meteorology is affected by wind speed and direction, atmospheric stability, inversion height and terrain features. Useful mathematical models must be able to account for all these parameters.

Air pollution models vary in complexity from simple microscale dispersion models, to sophisticated mesoscale, multi-source models describing transport, dispersion and photochemical reactions of pollutants.

Microscale models are used to estimate the ambient pollution levels near a single source or project, $C_{project}$. Mesoscale models are used to estimate the area-wide impact of a proposed source or project, or the background concentrations, $C_{background}$, due to other sources.

Ambient air pollution concentrations occurring downwind of a source consist of two components: pollution contributed directly by the source, and background. In most analyses, these components should be determined separately. The total air quality impact, represented in Figure 1 by the concentration C_x, is equal to the sum of the background plus the concentration contributed by the project under study. Whenever other major sources are nearby, their contribution of pollution must also be added to the project contribution. The objective of the air quality analysis is to determine the value of C_x as accurately as possible and to compare that concentration to the NAAQS (National Ambient Air Quality Standards).[2]

The role of monitoring is to measure ambient pollution concentrations, meteorological parameters and/or source strength parameters. Air monitoring at carefully selected sites provides a direct measurement of background concentrations. Measurements of meteorological or source strength parameters can be used to verify model input data or validate air pollution dispersion models. Validating microscale models requires source strength, meteorology, and both microscale and mesoscale (background) air pollution concentration measurements. Validating mesoscale models requires that source strength, meteorology, and air pollution measurements all be representative of area-wide conditions.

OBJECTIVES OF AIR MONITORING

Air quality monitoring is usually undertaken to characterize air quality in urban areas, near large point or line sources of pollution or where there are sensitive environmental receptors.

The ability to assess the air quality of an area depends on accurate data describing existing conditions and models which can be used to predict future pollution levels. Historically, two survey design methods have been used in community monitoring network design. The first is the area method in which spatial variations are handled by uniformly spaced stations with the hope of measuring averaged air quality. The other method consists of measuring air quality in the high pollution areas near point sources. The area method may be appropriate for flat terrain with few large point sources, while the source orientation method may be appropriate for hilly terrain. However, both of these types of network designs require extensive analysis if the data are to be evaluated for such things as seasonal variations and the contribution to the measured air quality of multiple sources.

The basic air monitoring site criteria for many air monitoring agencies require that each station measure an air mass that is representative of a relatively large land area (mesoscale). The actual location is based upon population density, industrial location and weather factors. Primary pollutant transport accompanied by photochemical reactions may require primary and secondary pollutant monitoring stations. Well-located stations avoid local sources of high pollutant concentrations (refineries, chemical plants, power plants, tank farms, major highways). The influence of these sources is determined by measuring the discharge of the pollutants into the atmosphere (stack monitoring) and utilizing short-term monitoring and modeling to determine the air quality near the sources (microscale). The local influence of buildings and hills must be considered as well as the location of trees and foliage, which may bias the local air quality concentrations. Stations located according to these criteria measure air quality where people live and work and disclose the effect of both large and small sources on air quality.

A preliminary air quality survey utilizing air quality modeling can be conducted as a means of selecting long-term sampling sites that will represent maximum pollutant concentrations or general exposure in a residential or commercial area. A well-designed survey would identify the variations in air quality in the area of interest and also show the extent of the area represented by the air quality measurements at the long-term air monitoring station.

More complex concepts of site selection require different types of stations. Primary and secondary stations can operate on different time scales to achieve optimized temporal and spatial resolution. Some stations will be located to acquire background (mesoscale) data while others will obtain information near sources (microscale). Different sampling methods and sample averaging times may prove to be cost effective for stations with different objectives. Real time meteorological and source strength monitoring will be required to interpret data collected at microscale stations where the data will be highly variable, depending on source strength and meteorological conditions. In addition, station density will be greater where air quality and meteorology is predicted to have a large spatial and temporal variation.

Obtaining Background Concentration Data

Background is defined as the pollutant concentration level which, when added to the contribution from a specific source and local contributions from other major sources, will give the total pollutant concentration level. Specific sources of air pollution include industrial, commercial, residential, institutional, highway and natural sources. Background concentrations can be determined either by monitoring or modeling. Mesoscale simulation models can be used to determine the pollutant level at any location, for the present or future, and under a wide range of conditions, while monitoring is conducted at a limited number of sites and the measured values are only those that occur for the conditions observed during sampling.

Model accuracy can be checked using monitoring data by comparing measured background levels to those predicted by the model for similar conditions. A sufficient number of background measurements should be made to determine long-term average air pollution concentrations, and also reflect fluctuations in background levels. Diurnal, daily, weekly and seasonal trends should be measured. Peak levels of background pollution are especially important. It is the peak or "worst case" background concentrations, added to the pollution levels predicted for a project, that determine the overall impact of the project. Background concentration measurements can also be compared to the meteorological conditions associated with peak background concentration measurements to provide insight into the cause of high air pollution concentrations.

Determining Compliance with Air Quality Standards

The federal government has set air quality standards for six air pollutants (Table 1). Each primary and secondary standard sets the air pollution concentration, which is not to be exceeded more than once a year. Primary standards are designed to protect public health, while secondary standards are designed to protect public welfare. Also specified in the air quality standard is the averaging time for which the maximum allowable concentration is applicable. For example, the standard for carbon monoxide is 35 ppm averaged for one hour, or 9 ppm averaged over eight hours. Measured values of carbon monoxide can be compared to the air quality standards only if they are averaged over a one-hour or an eight-hour sampling period as specified by the standard.

The number of measurements required to check for compliance with air quality standards is highly variable, depending on the averaging time of the standard and the predictability of the temporal distribution of pollutant concentrations. For example, short-time peak concentrations of some pollutants occur during predictable times of the day and/or seasons of the year. A check for compliance with short-time average pollutant concentrations may require monitoring only during these predictable

Table 1. Air Quality Standards

Pollutant	Type of Standard	Averaging Time	Frequency Parameter	Concentration	
				$\mu g/m^3$	ppm
Carbon monoxide	Primary and secondary	1 hr 8 hr	Annual maximum[a] Annual maximum	40,000 10,000	35 9
Hydrocarbons (nonmethane)	Primary and secondary	3 hr (6 to 9 a.m.)	Annual maximum	160[b]	0.24[b]
Nitrogen dioxide	Primary and secondary	1 yr	Arithmetic mean	100	0.05
Photochemical oxidants	Primary and secondary	1 hr	Annual maximum	160	0.08
Particulate matter	Primary	24 hr 24 hr	Annual maximum Annual geometric mean	260 75	– –
	Secondary	24 hr 24 hr	Annual maximum Annual geometric mean	150 60[c]	– –
Sulfur dioxide	Primary	24 hr 1 yr	Annual maximum Arithmetic mean	365 80	0.14 0.03
	Secondary	3 hr 24 hr 1 yr	Annual maximum Annual maximum Arithmetic mean	1,300 260[d] 60	0.5 0.1[d] 0.02

[a]Not to be exceeded more than once per year.

[b]As a guide in devising implementation plans for achieving oxidant standards.

[c]As a guide to be used in assessing implementation plans for achieving the annual maximum 24-hour standard.

[d]As a guide to be used in assessing implementation plans for achieving the annual arithmetic mean standard.

periods of high concentration. However, for air quality standards having an averaging time of one year, an air monitoring program would have to be designed that would include a sufficient number of air samples, taken throughout the period of a year, to be representative of the annual average pollutant concentration.

Validation, Calibration and Development of Models

Air quality monitoring investigations can be specifically designed for model validation and calibration. This type of air quality monitoring study is most appropriate under the following conditions:

1. When local topographic features violate the basic assumptions of the simulation model, field measurements can be used to "correct" the model for different types of terrain.
2. When no applicable mathematical model exists, field measurements can be used to develop an "empirical model," which can then be used to assess the air quality impact of the project.

A good mathematical model must be able to predict air pollution levels under changing conditions of source strength, wind speed and direction, atmospheric stability, and at various locations in both the microscale and mesoscale study area. A model's ability to predict pollution levels under changing conditions can only be tested after field measurements are taken under similarly changing conditions.

An air quality monitoring investigation designed to validate or calibrate an air quality model includes real-time field measurements of all the input variables to the model as well as the observed ambient air pollution concentration. Only by measuring all of the input variables, and accounting for all errors in the field measurements, can the remaining error be attributed to the inaccuracy of the model prediction. Air pollution measurements taken without source strength and meteorological measurements do not allow a quantitative interpretation of the different levels of air quality that are measured. Therefore, it will not be known whether observed high concentrations are due to uncontrollable (meteorology) or controllable factors. It is also important that a wide range of various field conditions be observed during the study. This requirement sometimes causes air monitoring studies to be conducted for rather long durations.

Determining Air Quality Trends

Once air pollution trends have been established and quantified, estimates of future pollution levels may be extrapolated. A valuable analytical method which can be used in conjunction with the extrapolation of trends is the correlation of the measurements during a short-term study with data from a long-term air monitoring station. The correlation can take the form of an empirical equation, determined using linear regression methods, or, it can simply indicate that air pollution concentrations measured within the project are always less than the concentrations observed at the historical station. Only when both monitoring stations are influenced by the same meteorology and similar pollution sources can the air quality measurements of the two stations be correlated. Air quality monitoring stations that are affected by different air pollution sources are likely to produce dissimilar results.

DESIGN OF AIR QUALITY SURVEYS

Source information on the type, location and rate of emission of various materials is fundamental to (a) the design of air quality surveys, (b) the interpretation of air quality data, and (c) the prediction of present and future concentrations of specific pollutants by mathematical models. The number and spatial distribution of sources can be described in two basic categories:

1. A point source is a large single emission point such as a power plant.
2. An area source is a number of small emissions distributed over a defined area such as residential home heating, or auto emissions on collector roads carrying low volumes of traffic.

An emission inventory that categorizes emissions into systematic groups provides data which can be used to characterize air quality. Emission factors, which are average rates of emissions for particular source types, make it possible to calculate average rates of emissions for point, line and area sources.

The transport and dilution of air pollutants is a function of wind speed, direction, air turbulence and topography. The effective dispersion of gaseous or fine material released into the atmosphere near the ground depends on natural mixing caused by turbulence. Turbulence depends on radiant energy received from the sun and this varies from day to night, with cloud cover, with the time of year and with location. The dynamic nature of air quality can thus be readily associated with the dynamic nature of meteorology.

Mathematical models are available that predict the change in concentration of material with atmospheric conditions. These models are based on idealized consideration of air flow and terrain, and provide theoretical estimates of the likely concentration levels and their temporal and spatial variation now and in the future. Figure 2 shows the basic relationship between idealized mathematical models for predicting air quality and surveys designed to measure air quality. There is feedback interaction between idealized models and air quality monitoring in which the models are used to provide idealized temporal and spatial variations in air quality. These idealized variations can be utilized in the design of air quality surveys, and the air quality data that are collected during a study can be utilized to validate or calibrate the idealized model. The corrected or refined model can then be used to generate a general profile of the temporal and spatial variations in the air quality of an area for both present and future conditions, based upon the finite amount of temporal and spatial monitoring data that were collected during the air quality survey.

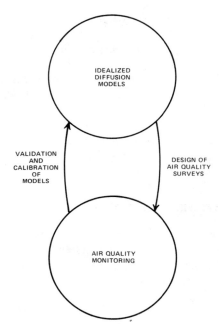

Figure 2. Interaction of modeling and monitoring.

The design of air quality surveys in this book is based on the premise that surveys that will identify the air quality of an area in the shortest possible time and at the least overall cost can only be designed by utilizing idealized mathematical models. The models provide specific information for use in site selection (degree of spatial resolution) and survey duration and sampling interval (degree of temporal resolution).

The second premise on which the air quality survey designs are based is that the major objective of the survey is to validate and/or calibrate an idealized mathematical model. Measurements of pollutant values and meteorological parameters are used to provide a refined mathematical model which then provides a practical method for calculating the general spatial and temporal air quality patterns near point, line and area sources.

In summary, the basic approach to the design of air quality surveys taken in this book is to utilize theoretical estimates of the likely concentration levels in order to reduce the number of locations requiring monitoring to those that are critical to a refinement of the idealized models or the development of an empirical model for important locations (*i.e.,* maximum expected concentrations near sensitive receptors). This method is used because it is assumed that it would require a prolonged sampling period at many locations to adequately characterize the dynamic nature of air quality in urban areas and near large point and line sources of pollutant by monitoring alone. Such surveys would be laborious and expensive.

Because of the cost of air monitoring sensors, it is critical that a minimum but sufficient number of stations be installed to produce timely data with the desired level of confidence. Thus, the primary consideration in survey design is to obtain the minimum number and location of sampling stations so that the data that is generated is adequate enough to calibrate and/or validate the idealized model which then allows generalized air quality estimates to be made for different combinations of source strength and meteorological conditions.

PROCEDURE FOR DESIGNING AN AIR QUALITY SURVEY

The design of an air quality survey is reviewed in Figure 3. The figure identifies the sequential nature of the decisions required to provide results that will meet the overall objective of the study.

The proper development of a cost effective air quality survey involves not only air quality monitoring, but also meteorological monitoring, calibration and data acquisition systems. Failure to recognize this fact at the outset results in a design based on many practical compromises that may fail to meet the sampling objectives. A cost effective system should reflect both the realities of current air quality monitoring system technology and the ultimate application for which the measurement system is intended. It is especially important to develop a systematic plan for the implementation of the system in advance of choosing specific pieces of hardware.

Air quality surveys require the determination of air quality in both the microscale and the mesoscale and characterization of both primary (directly emitted) and secondary (formed in the environment) pollutants. Sufficient temporal and spatial resolution of air quality must be obtained to meet the specific objectives of the survey within the broad objectives of characterization of air quality in urban areas, near sources and where there are sensitive receptors.

Set Objectives

The starting point of the air quality study is the setting of objectives for the air monitoring survey. Abstract objectives, like obtaining background

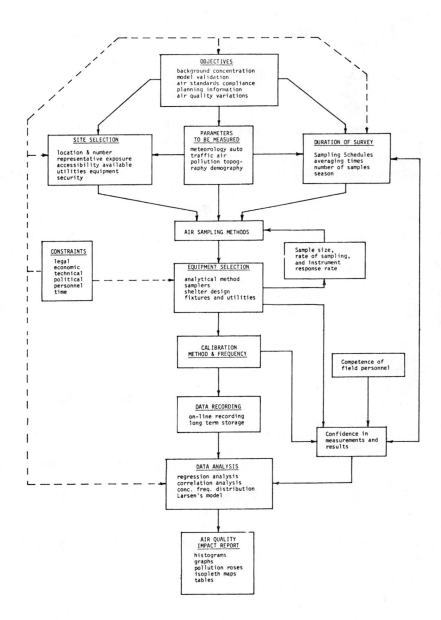

Figure 3. "Ten Step" procedure for designing an
air sampling survey.

levels or validating mathematical models, must be translated into clear meaningful statements in terms of the desired sampling results, such as the quantity of data needed, what air pollutants should be measured, the season of the year in which the sampling should be undertaken, and where and how many sampling sites should be chosen.

Choose Parameters to be Measured

Once the objectives are clearly defined, the next step is to determine the physical parameters to be measured. If the objective is to obtain background concentrations of pollution, then only the ambient levels of specific material will need to be measured. If the objective is to validate a mesoscale model so as to *predict* background concentrations, then meteorological conditions (*i.e.,* wind speed and direction and mixing depth) must be monitored in addition to the ambient levels of pollution. If the objective is to validate or calibrate a microscale model or develop an empirical model to predict pollutant concentrations, then real-time measurements of air pollution concentrations, meteorological conditions, and source strength conditions must be taken. The topographic and demographic features of the study area should also be determined and taken into consideration in the design of the field experiment.

Selecting Sampling Sites

Step three is the selection of the sampling sites. In choosing a technically suitable location for an air quality monitoring site, one must consider the representativeness of the site in terms of its exposure to air pollutants and prevailing meteorological conditions. Site exposure is heavily dependent on the relative location of pollutant sources and the effects of terrain on meteorological conditions. Specific station objectives should consider where the maximum background concentrations are expected to occur, and where the maximum project contribution to air pollution is expected to occur. The location of human receptors should be taken into account. It should be recognized that site selection is a very critical element in the survey design. If the wrong site is picked, or if a critical site is missed, no amount of accurate collection of data will allow the objectives of the study to be fully realized.

The meteorological exposure of an air monitoring site should be representative of either area-wide meteorology, or localized meteorological regimes that will effect the dispersion of air pollutants. Localized meteorological regimes result from irregular terrain features which can cause such things as increased mechanical turbulence or localized drainage winds.

The number of monitoring sites that should be used in the study will depend on the diversity of the land use patterns, meteorological regimes, source design configurations and sensitive receptors. This number will be limited by the number of technically suitable sites available, the constraints of money and manpower available and the level of statistical significance or confidence desired in the results.

Schedule Sampling

The fourth step is to determine the duration of the project and sampling schedules that are consistent with the objectives of the overall study. Air sampling schedules can be designed on a random or a systematic basis for both long- and short-term studies. In general, studies that are designed to validate simulation models are of a shorter duration than studies designed to document the extent of an existing pollution problem. In the case of photochemical oxidants, sampling only during the summer may be all that is needed. The availability of a historical air quality record within the urban area may allow another season to be utilized with data correlation providing the worst case conditions. If the objective is to collect data to validate a microscale pollution dispersion model, then this might be accomplished by conducting a very extensive monitoring investigation but having a duration of only one week.

Choosing Air Sampling Methods

The fifth step is to choose the type of air sampling method to be used. One can select continuous air monitoring, integrated grab sampling, intermittent sequential sampling or a combination of these. The air sampling method chosen depends on the air pollutant to be measured, the available utilities at the monitoring site, the frequency of sampling at the site, the number of sites, the distance between sites and the duration of monitoring required. In addition, the air sampling method must be compatible with the requirements of sample size, rate of sampling and the response rate of the analyzer to be used.

Selecting Equipment

The sixth step is the selection of the equipment to be used. The analytical method for measuring air pollutant concentrations must be chosen, a suitable instrument identified, and the equipment purchased. Air sampling and instrument calibration equipment must also be available. Air sampling shelters must be designed and fabricated.

Setting Calibration Procedures

Analyzers must be calibrated to insure accurate measurements. The calibration of an air pollution analyzer involves the preparation of gas mixtures of known air pollutant concentrations, which are injected into the analyzers, and the instrument response is adjusted to match the known pollutant concentration. There are basically two levels of sophistication in air monitoring instrument calibration. A primary instrument calibration requires that reference methods (generally wet chemical methods) of measurement be conducted in side-by-side operation with the analyzer being calibrated. A primary calibration will include a check of the instrument response at several different pollutant concentration levels (i.e., 20%, 40%, 60% and 80% of full scale) so that the instrument linearity can be checked. In general, however, a primary calibration is difficult to perform under field conditions requiring wet laboratory facilities. For this reason, primary calibrations are usually performed periodically (every 90 days) and supplemented by secondary calibrations.

Secondary calibrations depend on the generation of gas mixtures of known pollutant concentrations which can be used to check the analyzer response at a single upscale point. The output of the analyzer can then be compared for accuracy with the known gas concentration. In this way the instrument can be checked frequently to see if it is operating correctly. Gas mixtures of known pollutant concentration can be purchased in pressurized steel cylinders, or can be generated using permeation tubes. Ozone concentrations can be produced using an ultraviolet lamp such as an ozone generator. The frequency of calibration must be determined after observing the operating characteristics of the instrument under field conditions. Very stable instruments may need secondary calibration only several times each week, while unstable instruments may require secondary calibration several times per day.

Choosing Data Recording Methods

The eighth step is the selection of a suitable method of recording the field measurements. The recording methods available include (a) continuous strip chart recorders, (b) scanning analog or digital electronic data loggers, and (c) manual data recording by the instrument operator. The data recording method used depends primarily on the type of air sampling method employed. When using continuous analyzers, strip chart recorders or scanning data loggers are usually used. When using the intermittent sequential sampling method, or when analyzing grab samples, manual recording of the data on specially prepared data forms is frequently the most efficient and least expensive method. The engineer must also choose between on-site data recording and use of a telemetry system.

The important concern in data recording is that each air quality measurement be correctly time-referenced. When monitoring is conducted at several locations simultaneously, the data recorders used at each station must be carefully synchronized, so that a correct time reference can be assigned to each measurement when the data are analyzed. In general, it is a good procedure to analyze and record all measurements in the chronological order in which they are taken and to record instrument calibration data along with the air sample measurements.

Reducing air quality data is the process of analyzing the instrument monitoring records, comparing these records to the calibration results, and determining the actual parts per million of the pollution concentrations measured. Large volumes of data are usually generated. For example, a CO monitoring study with six monitoring stations operating 12 hours a day, four days a week, will generate over a thousand one-hour average CO measurements each month. Therefore, a simple and logical format should be developed for storage of the data, so that it can be easily retrieved and sorted when it is necessary to evaluate the entire data bank and prepare the final report.

Analyzing Data

Air quality studies inherently involve the taking of a limited number of samples from a very variable and uncontrolled population (the environment). For this reason, air quality data is analyzed using statistical methods. Special methods are available for the analysis of air quality data, which predict the behavior of the total population based on a limited number of samples. Limited field measurements may not record the maximum concentrations, but statistical analysis can be used to determine these values.

The types of air quality data analysis that are very useful in air quality assessment include (a) concentration frequency distribution, (b) averaging time analysis, (c) spatial distributions of pollutants, (d) regression analysis and (e) determination of compliance with NAAQS.

The magnitude of the error due to the sampling procedure is governed by the type of analyzer used, the method and frequency of calibration, and the competence of the field personnel. The statistical significance of the results is strongly dependent on the sampling frequency and the duration of the study.

Report Results

The final step is to determine how to present the results. Results of the monitoring can be presented in the form of graphs, histograms, pollution roses and isopleth maps. The method chosen should be the one that best illustrates the relationship of an observed pollutant concentration with some other variable. Graphs of pollution concentration versus time are perhaps the best way to illustrate diurnal, daily and seasonal trends in pollution level. The frequency of occurrence of various pollution levels can best be illustrated by plotting the cumulative frequency distribution of pollution concentrations on log probability graph paper. When the pollution concentration is dependent on the wind direction, pollution roses can be used to illustrate the relationship. In most cases, the observed pollution concentrations vary at different geographical locations. In this case, isopleth maps showing the average pollution concentrations at different locations represent a good method for displaying the results of the air quality study.

REFERENCES

1.   National Environmental Policy Act, U.S.C., Public Law No. 91-190 (1970).
2.   "National Primary and Secondary Air Quality Standards," *Federal Register*, 36:84 (April 30, 1971).

CHAPTER II

MATHEMATICAL MODELS FOR PREDICTING AIR QUALITY

INTRODUCTION

The abiltiy to predict air quality by the use of mathematical models usually involves a number of simplifying assumptions about the dispersion of pollution. This produces models that are useful but not highly accurate.[1] The present state-of-the-art for these models usually involves the use of generalized solutions that allow relative concentrations to be determined for different conditions of atmospheric mixing and distance from the source-emitting material. This section will review some of the models that are used for point, line and area source air quality predictions and show the general solutions that can be used to obtain pollution concentration variations with time and location.

NONGAUSSIAN ATMOSPHERIC DIFFUSION MODELS

Box Model

The simplest form of a dispersion model is the "box" model, which is often used to estimate air pollution concentrations due to area sources. The "box" itself represents the three-dimensional volume within which air pollutants are assumed to be thoroughly mixed. The box is usually defined as having unit width and oriented so that its length, S, lies in the direction of the mean wind, U. Pollutants are assumed to be emitted at a constant rate per unit time per unit area, Q (gm/m^2-sec). Z equals the height of the box, to which pollutants are dispersed (see Figure 4). QS is equal to the emission rate per unit width, which when divided by the ventilation rate, UZ, is equal to the equilibrium concentration, C_e, at a point S distant from the upwind edge of the area source.

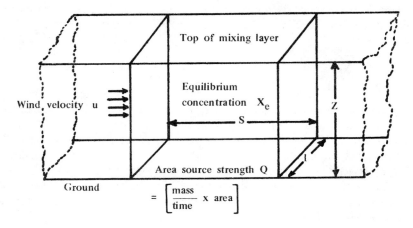

Figure 4. Area-source "box" diffusion model.[2]

$$C_e = \frac{QS}{UZ} \tag{1}$$

Box models can be used to obtain order-of-magnitude estimates of ambient pollution levels; however, the simplifying assumptions of the model (uniform mixing to a constant level, Z) lead to results that do not simulate true atmospheric conditions accurately.[2] Special application of the box model is an elevated inversion above a point source in a valley. Then,

$$C = \frac{Q}{HWU} \tag{2}$$

where W = the width of the valley
       H = the mixing height

K-Theory

Some atmospheric dispersion models have been developed that are theoretically based on diffusion theory. According to Fick's law, the rate of diffusion of a substance in solution (*i.e.,* a pollutant in air), q, from a high concentration to a lower concentration is proportional to the concentration gradient: (in the X-direction)

$$\frac{dq}{dt} = K \frac{\partial^2 q}{\partial x^2} \tag{3}$$

where K is constant and equal to the diffusivity or diffusion coefficient.[3] For atmospheric diffusion the equation may be written for three dimensions:

$$\frac{dq}{dt} = \frac{\partial}{\partial x}(K_x \frac{\partial q}{\partial x}) + \frac{\partial}{\partial y}(K_y \frac{\partial q}{\partial y}) + \frac{\partial}{\partial z}(K_z \frac{\partial q}{\partial z}) \tag{4}$$

K values are assumed to be determined solely by atmospheric properties such that the rates of diffusion of heat, K_h, turbulent momentum, K_m, and pollutants K_p in the atmosphere are related. According to Hanna,[4] during adiabatic conditions $K_h \cong K_m \cong K_p$. During unstable conditions $K_p \cong 3K_m$.

The most satisfactory application of K-theory has been to the case of crosswind dispersion from an infinite continuous line source. For a mean wind $\bar{u}$ on the X-axis, it is a reasonable assumption that:

$$\bar{u} \frac{\partial q}{\partial x} \gg \frac{\partial}{\partial x}(K_x \frac{\partial q}{\partial x}) \tag{5}$$

for which a solution has been given by Hanna.[4]

$$\frac{q}{Q} \cong \frac{1}{(2\pi K_z x \bar{u})^{\frac{1}{2}}} \exp(-\frac{\bar{u} \ z^2}{4K_z x}) \tag{6}$$

According to Hanna,[4] K_z can be calculated as a function of the friction wind speed, u_*, and height, Z:

$$K_z = 0.4 \ u_* \ Z \qquad 0 < Z < 100 \ m \tag{7}$$

$$K_z = (40m)(u_*) \qquad 100m < Z < 100m \tag{8}$$

In practice K values are seldom determined from heat flux or turbulence flux measurements due to the difficulty of making these measurements. Rather, they are determined empirically from observed pollutant diffusion data. Also, the assumption that pollutant diffusion is Fickian (constant K) cannot be valid because of the known variations of mean wind shear and heat flux with time and space. Hence, though the basic equations are founded in diffusion theory, K-theory is at best semi-empirical.

GAUSSIAN PLUME MODELS

Point Sources

The most widely employed method of modeling air pollution dispersion is the use of Gaussian plume equations. These equations can be derived from a simple dimensional analysis plus the basic Gaussian distribution function. Air pollutants emitted at a constant rate, Q(gm/sec), from a smokestack are commonly referred to as a "point source" from which pollutants are released into the atmosphere at a "point" in space located at the top of the stack. The material is then transported by the wind away from the source at a rate equal to the wind speed, U(m/sec), at the point of release. Hence, the pollutants are distributed in the down-wind direction, X, with a mass density equal to Q/U (gm/m) (see Figure 5). Thus, for a conservative pollutant (no chemical reactions or fallout), the mass of pollutant contained in any volume of the plume with incre-mental length, ΔX, is equal at all distances downwind. However, the concentration C(gm/m^3) within the plume decreases with increasing dis-tance downwind because atmospheric turbulence tends to disperse the material in the horizontal (y-direction) and vertical (Z-direction) such that the plume "spreads" about its centerline. The average concentration

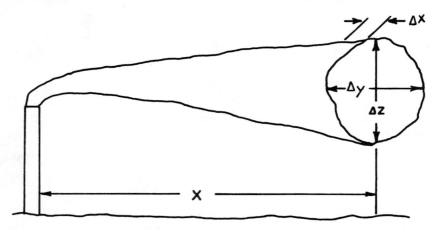

Figure 5. Dispersion of the plume from a point source (X = downwind distance, Δx = incremental distance, Δy = plume width and Δz = plume height).

of pollutant within the plume at a point, X, downwind is then inversely proportional to the amount of spreading of the plume (Δz and Δy) and the transport wind speed, U:

$$C \propto \frac{Q}{U\ \Delta z\ \Delta y} \tag{9}$$

From observations it has been found that the concentration within the plume is not uniform, but is highest near the center and decreases toward the outer edges of the plume. Both experimental results and theory[1] indicate that the concentration profile within a plume follows a bell-shaped normal or Gaussian distribution curve (see Figure 6) where the peak of the curve (the population mean) describes C near the plume centerline and a point ± 3 standard deviations from the mean approximates the outer edge of the plume. The Gaussian distribution function for an x mean equal to zero, and standard deviation, σ is:

$$Y = \frac{1}{\sigma(2\pi)^{\frac{1}{2}}} \exp\left(-\frac{x^2}{2\sigma^2}\right) \qquad (10)$$

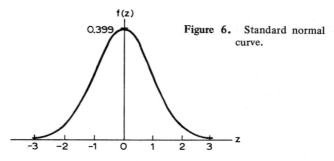

Figure 6. Standard normal curve.

In practice, σ equals the distance from the plume centerline to the point where the concentration is 0.6066 of the centerline concentration. The Gaussian function for both the Y and Z directions can then be used to replace $\Delta y = \sigma_y(2\pi)^{\frac{1}{2}}$ and $\Delta z = \sigma_z(2\pi)^{\frac{1}{2}}$, yielding,

$$C = \frac{Q}{2\pi\, U\, \sigma_y\, \sigma_y} \exp - \frac{1}{2}\left[\left(\frac{y}{\sigma_y}\right)^2 + \left(\frac{z}{\sigma_z}\right)^2\right] \qquad (11)$$

where y and z equal the distance from the plume centerline. The most common convention is to use the coordinate system illustrated in Figure 7 such that Equation 11 is written:

$$C(x,y,z,H) = \frac{Q}{2\pi\, U\, \sigma_y\, \sigma_z} \exp - \frac{1}{2}\left[\left(\frac{y}{\sigma_y}\right)^2 + \left(\frac{H_e\text{-}Z}{\sigma_z}\right)^2\right] \qquad (12)$$

where H_e = the effective stack height. Further, beyond the point, x, where the plume first touches the ground the plume is assumed to be "reflected" because the earth's surface is a barrier to further dispersion. If the pollutant is not deposited or absorbed, then ground-level concentrations downwind of the touchdown point are underestimated by Equation 12. To account for "reflection" of the plume a mirror image

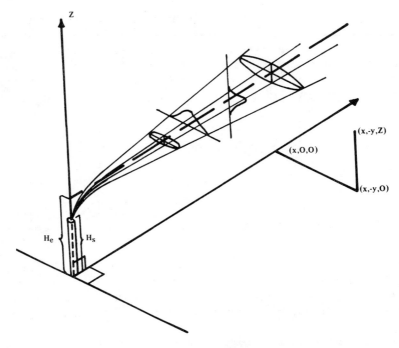

Figure 7. Coordinate system showing Gaussian distributions in the horizontal and vertical.

source is assumed to exist, located symmetrically (with respect to the ground) to the actual source (see Figure 8). The final form of the Gaussian plume equation for a continuous point source is then,

$$C(x,y,z,H) = \frac{Q}{2\pi \ U \ \sigma_y \ \sigma_z} \ exp - \frac{1}{2} (\frac{y}{\sigma_y})^2 \quad exp \ (-\frac{1}{2} (\frac{z-H}{\sigma_z})^2)$$

$$+ \ exp \ (-\frac{1}{2} (\frac{z+H}{\sigma_y})^2) \tag{13}$$

Diffusion Coefficients σ_y and σ_z

The determination of σ_y and σ_z for Equation 13 is a crucial step in predicting ambient pollution levels using the Gaussian model. Various authors (i.e., Smith,[5] Pasquill,[6] Brookhaven,[7] Briggs[8] and TVA[9]) have summarized the results of dispersion studies and presented empirical values

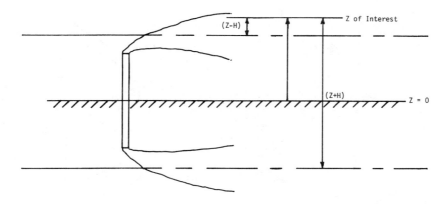

Figure 8. Coordinate system of real source and imaginary source.

for σ_y and σ_z for use in Gaussian models. Each of these studies points out the dependence of diffusion rates on the turbulent condition of the atmosphere, which varies both temporally and spatially. Temporal variations due to changing insolation, vertical temperature structure and wind speeds result in different values of σ_y and σ_z during different times of day, etc. For this reason empirical values for σ_y and σ_z have been categorized by Pasquill[6] and others according to defined atmospheric stability classes (see Table 2). Pasquill's diffusion coefficients can then be determined using Figures 9 and 10 according to the stability class and downwind distance.

Table 2. Pasquill's Atmospheric Stability Categories

| Surface Wind Speed (m/sec) | Insolation | | | Night | |
| | | | | Thinly Overcast or | |
	Strong	Moderate	Slight	$\geqslant$4/8 Low Cloud	$\leqslant$ 3/8
<2	A	A-B	B	–	–
2-3	A-B	B	C	E	F
3-5	B	B-C	C	D	E
5-6	C	C-D	D	D	D
>6	C	D	D	D	D

Key to Stability Categories

For A-B take average of values for A and B, etc.

Strong insolation corresponds to sunny midday in midsummer in England, slight insolation to similar conditions in midwinter. Night refers to the period from one hour before sunset to one hour after dawn. The neutral category D should also be used, regardless of wind speed, for overcast conditions during the hour preceding or following night as defined above.

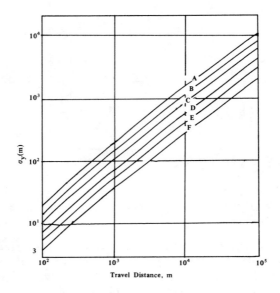

Figure 9. Horizontal diffusion coefficient versus travel distance.

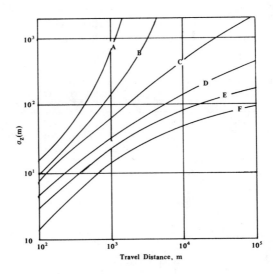

Figure 10. Vertical diffusion coefficient versus travel distance for Pasquill's turbulence types.

Spatial variations in atmospheric turbulence occur due to the effects of wind shear with height and surface roughness. In general, turbulence intensity is greater near the earth's surface than at elevated heights (100s of meters) and it is generally greater over rough terrain (hills, buildings, etc.) than over smooth (grassland or desert). Hence, different diffusion coefficients may be needed for modeling very tall stacks or, conversely, diffusion in downtown areas, because Pasquill's curves are based on ground source data with smooth terrain.

Briggs[8] suggests that different values of σ_y and σ_z be used for urban and rural terrain because diffusion is more rapid in urban surroundings due to the mechanical turbulence produced by wind flow over buildings and the convective turbulence generated by loss of building heat. Diffusion coefficients suggested by Briggs are given in Figures 11-14.

TVA[9] and Hanford[10] have published the results of diffusion studies on tall stacks which indicate that for stable conditions the use of Pasquill's diffusion coefficients may result in as much as an order-of-magnitude underestimation of the axial plume concentration at distances of 10 km, according to Islitzer.[11]   TVA's curves for σ_y and σ_z are shown in Figures 15 and 16.

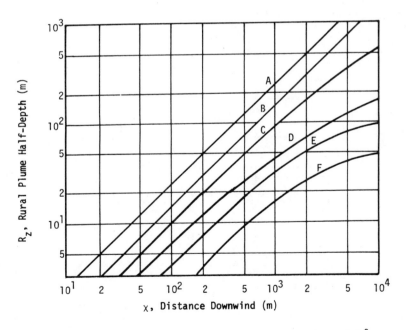

Figure 11. Rural plume half-depth, R_z, versus distance downwind.[8]

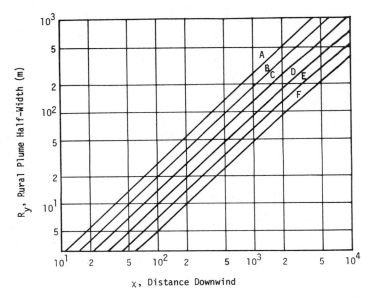

Figure 12. Rural plume half-width, R_y, versus distance downwind.[8]

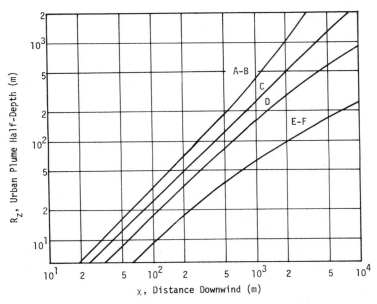

Figure 13. Urban plume half-depth, R_z, versus distance downwind.[8]

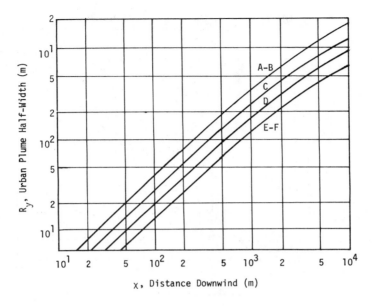

Figure 14.   Urban plume half-width, R_y, versus distance downwind.[8]

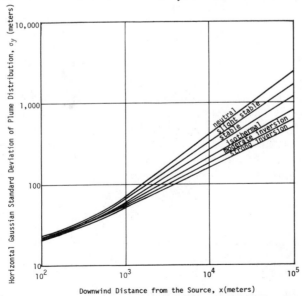

Figure 15.   Horizontal Gaussian standard deviation of plume distribution as a function of downwind distance from the source.[9]

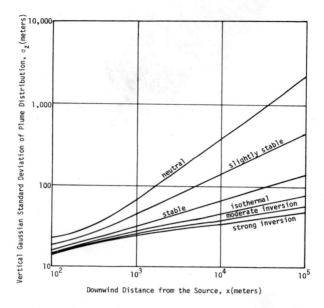

Figure 16. Vertical Gaussian standard deviation of plume distribution as a function of downwind distance from the source.[9]

Line Sources

Line source models are used to simulate dispersion from highways where cars are continually emitting pollutants. The form of the Gaussian plume equation used depends on the configuration of the highway and the wind direction. The simplest case is for a ground level "infinite" line source and perpendicular (crosswind) conditions when pollutants are transported in the X direction and dispersed only in the Z direction. Turner[12] suggests the use of the equation:

$$C(x,z) = \frac{2 \, Q_L}{(2\pi)^{1/2} \, \sigma_z \, u} \, \exp \, -\frac{1}{2} \left(\frac{z}{\sigma_z}\right)^2 \tag{14}$$

where Q_L = emission rate per unit length of highway (gm-sec^{-1}-m^{-1})

x = distance separating road and receptor

z = height of receptor above ground.

This equation results in a downwind concentration profile such as illustrated in Figure 17. The condition that the line source be infinite in

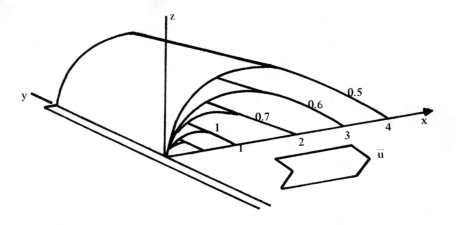

Figure 17. Cross section revealing surface of constant ambient concentration for an infinitely long line source on the y-axis. Only relative concentrations are indicated.

extent does not introduce large errors as long as the finite road length is greater than $3 \sigma_y$ long in both directions from the receptor. If the road is shorter, then the "edge effects" caused by the end of the line source must be accounted for. These edge effects will significantly increase at greater distances from the source. Sutton[13] presented the following equation in 1932 for application to finite line sources:

$$C(x,y,0) = \frac{2 \, Q_L}{(2\pi)^{\frac{1}{2}} \, \sigma_z \, \bar{u}} \, \exp -\frac{1}{2} (\frac{z}{\sigma_z})^2 \int_{P_1}^{P_2} \frac{1}{(2\pi)^{\frac{1}{2}}} \, \exp -(\frac{p^2}{2}) \, dp \qquad (15)$$

where $P_1 = y_1/\sigma_y$ and $P_2 = y_2/\sigma_y$.

In order to account for different wind directions, ϕ (referred to the road), Turner[12] suggests a simple trigonometric correction ($1/\sin \phi \times$ Equation 14) which accounts for the decreased wind speed, u, normal to the road. Turner further recommends that this correction be confined to wind angles $>45°$. It should also be recognized that the distance separating the road and receptor, x, is increased under oblique wind conditions by a similar amount ($1/\sin \phi$) such that a different value of $\sigma_z(x)$ is required.

The most difficult case for modeling a line source is under parallel wind conditions when pollutants are transported in the x direction and dispersed in both the y and z directions. In this case a receptor located at point P in Figure 18 will be exposed to pollutant levels that result from the accumulation that occurs as winds pick up and transport

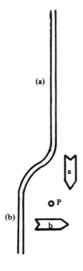

(a)

(b)

Figure 18. A line source curves around a receptor at P. Wind parallel to the long stretch of road (a) will produce a higher ambient concentration than transverse (b) provided that atmospheric mixing is minimal.

pollutants from the long highway section "a" illustrated in the figure. An equation can be derived for the simplest case of ground level source and receptor by integrating the contribution of each incremental length of highway, Δx, from the initial, x_1, and final, x_2, points at which the highway becomes parallel to the wind. This equation is:

$$C(x,o,o,o) = \int_{X_1}^{X_2} \frac{Q_L \, dx}{\pi U \, \sigma_z(x) \, \sigma_y(x)} \qquad (16)$$

if the diffusion coefficients $\sigma_z(x)$ and $\sigma_y(x)$ can be expressed in exponential form: $\sigma_z(x) = ax^p$ and $\sigma_y(x) = bx^q$, then the integral can be solved.

$$C(x,o,o,o) = \frac{Q_L}{\pi U} \int_{X_1}^{X_2} \frac{dx}{ab \, x^{(p+q)}} = \frac{Q_L}{ab \, \pi U} \frac{x^{1-(p+q)}}{1-(p+q)} \Bigg|_{X_1}^{X_2} \qquad (17)$$

Beaton et al.[14] have developed an empirical approach to line sources under parallel wind conditions. The equation for a ground level source and receptor is:

$$C(x,y,o,o) = A\left(\frac{Q}{U}\right)\left(\frac{1}{K}\right)\left(\frac{30.5}{W}\right) \exp - \frac{1}{2}\left(\frac{y}{\sigma_y}\right)^2 \qquad (18)$$

where   Q = the emission rate in gm/sec for 100 feet of highway
         K = 4.24
         W = width of road shoulder to shoulder (meters)
         A = an empirical coefficient which accounts for the accumulation of
               pollutants under parallel wind conditions. A is a function of road
               length and stability class (see Figure 19).

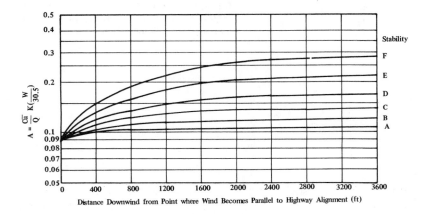

Figure 19. Ground level concentration ratio, A, downwind from highway line source
parallel wind at-grade section for all stability classes.[14]

Another approach which is taken by the EPA[15] for modeling line sources (*i.e.,* the EPA "Hiway" model) is to simulate a line source with a number of point sources each with an equal proportion of the total highway pollution emission rate. The sum of the contributions of each point source to air quality at the receptor, P, is equal to the contribution of the line source (see Figure 20). The advantage of this method is that the same point source equations can be used regardless of wind direction.

Diffusion Coefficients for Line Sources

As in the case of point sources, the diffusion coefficients for line sources have been determined empirically from field experiments. At present the two sets of diffusion coefficients most widely used are those employed by the California Division of Highways[14] and those used by EPA.[15] Figures 21 and 22 show the sigma curves used by California. Gilbert and Davis[16] have compared the σ_z curves used by California and EPA models. Figure 23 illustrates the differences in σ_z between the two models.

Figure 20. Illustration of how EPA "Hiway" model simulates a line source with multiple point sources.

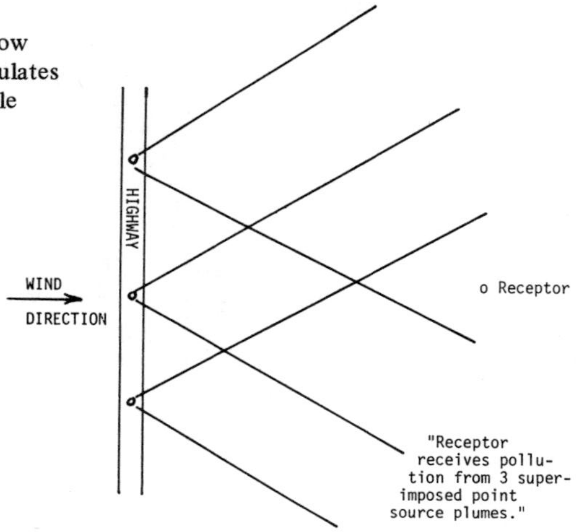

WIND DIRECTION

HIGHWAY

o Receptor

"Receptor receives pollution from 3 superimposed point source plumes."

Area Sources

Area source emissions is the term used in reference to the combined effect of the emissions of large numbers of relatively small sources of pollutant emissions such as residential, commercial and small industrial fuel-burning sources. The distribution of area source emissions is frequently assumed to be uniform over the urban area. The contribution of area sources to the pollutant level measured at an air quality station is a function of the emission density (μg m^{-2} sec^{-1}) and the upwind extent of the area source. The extent of the area source is defined as the distance from the air monitoring station, to the upwind edge of the area source within which the station is located. Figure 24 illustrates how the extent of the area source may change with direction. In this example the area source has an extent of 1-1/2 km to the east and 5 km to the northwest of the air quality station. The larger the size (or extent) of the area source, the higher the concentration; however, the effect of air quality of increasing area source size beyond 3 km is fairly small. This effect is illustrated in Figure 25. The technique used to determine the ground level pollutant concentrations resulting from area source emissions is similar to the technique employed by Miller and Holzworth.[17]

Figure 25 is the result of the integration of the Gaussian plume equation for a uniform area source extending upwind of a receptor any distance from 100 meters to 10 kilometers. The distance X is located on the abscissa, and the normalized concentration, CU/Q is located on the ordinate

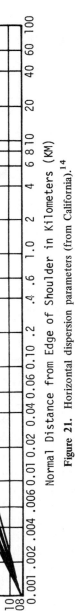

Figure 21. Horizontal dispersion parameters (from California).[14]

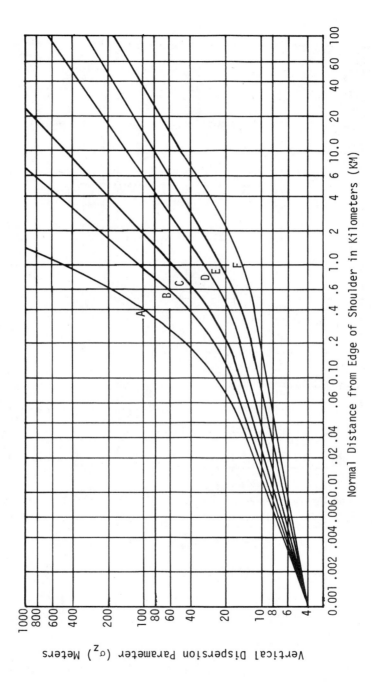

Figure 22. Vertical dispersion parameters (from California).[14]

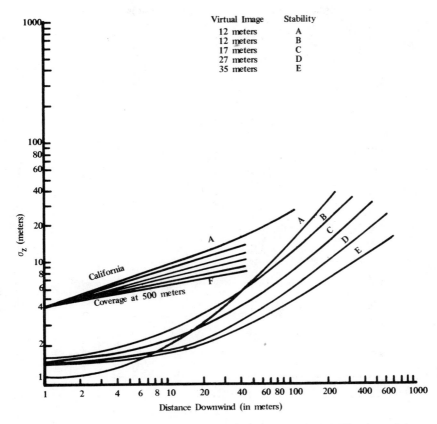

Figure 23. Comparison of σ_z values used in the EPA and California models.

axis. The Gaussian equation approximately representing area sources emitting at ground level is:

$$C = \int_{100m}^{X} 2Q \; dx/(2\pi)^{\frac{1}{2}} \; U \; \sigma_z(x) \qquad (19)$$

where C = concentration at receptor, $\mu g/m^3$
$\qquad\;\; Q$ = emission density, $\mu g/m^2 .s$
$\qquad\;\; x$ = upwind extent of area source from receptor, m
$\qquad\;\; U$ = wind speed, m/s
$\quad \sigma_z(x)$ = vertical dispersion parameter which is a function of distance from the source and atmospheric stability. In solving Equation 19

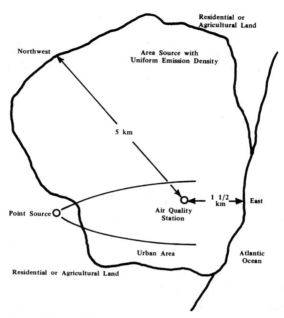

Figure 24. Determining the extent of the area source.

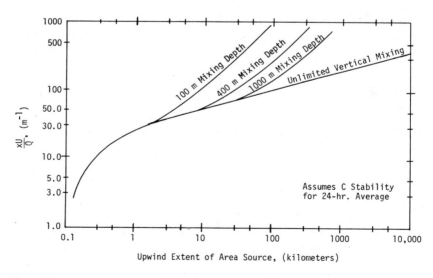

Figure 25. The effect on ground level pollutant concentration of increasing the size of the urban area source for various mixing depth.

a value for $\sigma_z = 0.12 \ x^{0.915}$ meters was used as determined by Calder.[18] This expression for σ_z assumes that C stability is the most applicable for calculating 24-hour average pollutant concentrations. Stabilities ranging from B to E occur almost every day with durations of minutes to many hours. Stability C is assumed to average out the effect of much of the early morning and nighttime stable conditions combined with the generally unstable conditions occurring during the middle of the day.

Given the highly approximate model underlying Figure 25, it can be seen that the effect of increasing the extent of the area source from 3 km to 30 km results in an increase in ground level concentration by a factor of 2. Under conditions of limited vertical mixing, limited by a mixing "lid," the concentration may be increased by a factor of 3 for mixing depths of 400 m and a 100 km area source. In general, varying area source size over a normal range of 3 to 10 kilometers will have a fairly small effect on pollutant concentration under all but the lowest depths.

The Gaussian model for area sources is one of the simplest forms of the Gaussian plume equation, yet urban areas represent one of the most complex sources for which modeling is desired. Hence, the simple calculations described below should not be expected to give precise results due primarily to the simplifying assumptions regarding the homogeneity of source strength throughout the urban area that is common to many area source models. To improve the ability of area source models to resolve more accurately the actual ambient concentrations, some models divide the urban area into grid squares or uniform blocks each having homogeneous source strength within, but different source strengths among the blocks. This type of modeling is exemplified by Gifford and Hanna's[19] simplified models or SRI-APRAC-1A[20] urban area source model. In the case of SRI's model,[20] a high-speed digital computer is needed to perform the many mathematical calculations.

PLUME RISE

Introduction

The height to which a plume of flue gases rises into the atmosphere after being emitted from a stack has a significant effect on the resulting ground level pollution concentration downwind. The Gaussian plume equations discussed previously require the effective stack height, H_e, as input information. H_e is equal to the sum of the physical stack height, h_s, and the plume rise, Δh.

$$H_e = h_s + \Delta h \tag{20}$$

Plume rise has been studied by numerous authors,[1,5,21-27] each publishing various empirical equations describing their results. Most authors generally recognize that plume rise is dependent on the initial momentum of the gases, the buoyancy of the gases (if at temperatures other than ambient), the vertical temperature structure of the atmosphere and the horizontal wind speed. One of the best consolidations of plume rise results and appropriate equations has been prepared by Briggs.[21]

Momentum Sources

Sources at temperatures close to the ambient and with significant exit speeds may be treated as jets in a crosswind. For this case Briggs[21] presents the results of Rupp et al.[24]:

$$\Delta h = 1.5 \ (\frac{V_s}{U})D \ = \ 1.5RD \tag{21}$$

where $R = V_s/U$, V_s is the stack exit velocity, U the average wind speed, and D the stack diameter. Briggs[21] recommends using the constant 3.0 in place of 1.5. Smith[5] gives a similar equation:

$$\Delta h = D(\frac{V_s}{U})^{1.4} \tag{22}$$

Jet rise to its final height is a function of the downwind position x and appears to be $(x/D)^{1/3}$ dependent. An expression for determining Δh as a function of x is given by Briggs[25] as

$$\frac{\Delta h}{D} = 1.89 \ (\frac{R}{1+3/R})^{2/3} \ (\frac{x}{D})^{1/3} \tag{23}$$

Buoyant Plumes

In general, the form of the plume rise equation for buoyant plume is

$$\Delta h \ \alpha \ (\frac{F}{US})^{1/3} \tag{24}$$

where F is the buoyant momentum flux parameter, U is wind speed, and S is an atmospheric stability parameter.

$$F = g \ V_s \ (\frac{D}{2})^2 \ (\frac{T_s - T_a}{T_a}) \tag{25}$$

$$S = \frac{g}{T} \ \frac{\partial \theta}{\partial z} \tag{26}$$

where g = acceleration of gravity
   T_s = stack temperature
   T_a = ambient temperature
   $\dfrac{\partial \theta}{\partial z}$ = vertical potential temperature gradient
   T = local potential temperature.

Experimental data have been evaluated and empirical equations for calculating plume rise have been presented by Briggs.[21]   For transitional rise:

$$\Delta h = 1.6 \; F^{1/3} \; u^{-1} \; x^{2/3} \tag{27}$$

For final rise under neutral conditions:

$$\Delta h = 1.6 \; F^{1/3} \; u^{-1} \; (10 h s)^{2/3} \; \text{for } h_s > 100 m \tag{28}$$

and

$$\Delta h = 1.6 \, F^{1/3} \quad u^{-1} \; (3.5 x*)^{2/3} \; \text{for } h_s < 100 m \tag{29}$$

where

$$x* = 0.52 \; \left(\frac{\sec^{6/5}}{ft^{6/5}}\right) \; F^{2/5} \; h_s^{3/5} \tag{30}$$

For final rise under stable conditions with wind:

$$\Delta h = 2.9 \left(\frac{F}{US}\right)^{1/3} \tag{31}$$

For final rise under stable and calm conditions:

$$\Delta h = 5.0 \; F^{1/4} \; S^{-3/8} \tag{32}$$

No reliable methods have thus far been proposed for calculating plume rise under unstable conditions.[26]

A Nomographic Solution for Plume Rise

From plume rise experiments TVA[27] has formulated an equation similar to Brigg's equation. TVA's equation is:

$$\Delta h = 114 \; C \; U^{-1} \; F^{1/3} \tag{33}$$

where C is an empirical coefficient equal to

$$C = 1.58 - 0.414 \frac{\Delta \theta}{\Delta z} . \tag{34}$$

The number 114 has dimensions $(m^{2/3})$.

Using this equation and mean values of $C = 1.58$ for neutral conditions and $C = 1.10$ for stable conditions, TVA[27] has prepared a nomogram, which can be used to provide rapid solutions of the plume rise (Equation 33). The nomogram is presented in Figure 26.

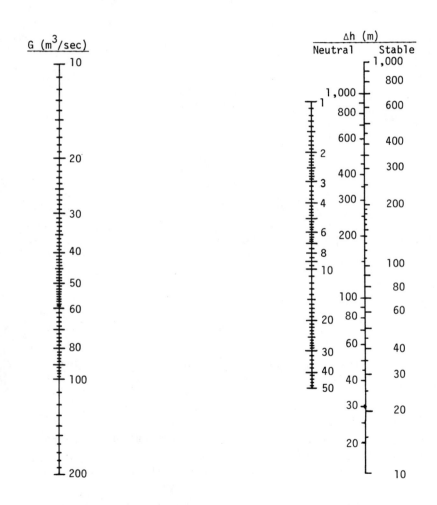

Figure 26. Buoyant plume rise, Δh, versus G and mean wind speed, $\bar{u}$, for neutral and stable atmospheric stability (from TVA).[27]

To use the nomogram, first calculate the parameter, G, where

$$G = V_s \, r^2 \, (p_a\text{-}p_s)/p_z \; (m^3/sec) \qquad (35)$$

and V_s = stack gas exit velocity (m/sec)
  r = inside radius of stack top (m)
  p_a = density of ambient air at stack top (gm/m^3), and
  p_s = density of stack gas at stack top (gm/m^3).

Then draw a straight line between the parameter G on the left scale and the mean wind speed at the plume centerline on the center scale. The calculated value for plume rise in either neutral or stable conditions can be read directly off the right hand scale.

GENERAL SOLUTIONS OF POINT SOURCE DIFFUSION MODELS

Introduction

To gain insight into the nature of the atmospheric diffusion problem, it is desirable to obtain general solutions to the models that point out the various relationships among input parameters, and how changes in these parameters affect downwind ground level pollution concentrations. Two different types of solutions are presented in this section: (a) the graphical solutions of Turner,[12] and (b) nomographic solutions by TVA.[27-29] These solutions are subsequently used for air monitoring site selection.

Graph Solutions

Turner[12] presents a convenient procedure for determining ground level plume centerline concentrations versus distance downwind from a point source. Turner plots the normalized pollutant concentration, $\chi u/Q$, against distance downwind for various effective stack heights, H, and stability classes (A-F); where χ = pollutant concentration (g/m^3), U = mean wind speed (m/sec), and Q = pollutant emission rate (g/sec). An example of this type of plot is shown in Figure 27 for C stability. Estimates of actual concentrations may be determined by multiplying ordinate values by Q/U.

Turner has also plotted normalized ground level concentration isopleths for both ground level sources and elevated sources (*i.e.,* H = 100 meters). Figures 28 and 29 illustrate the theoretical concentrations, $\chi u/Q$, as a function of downwind distance, x, and crosswind distance, y, for C stability.

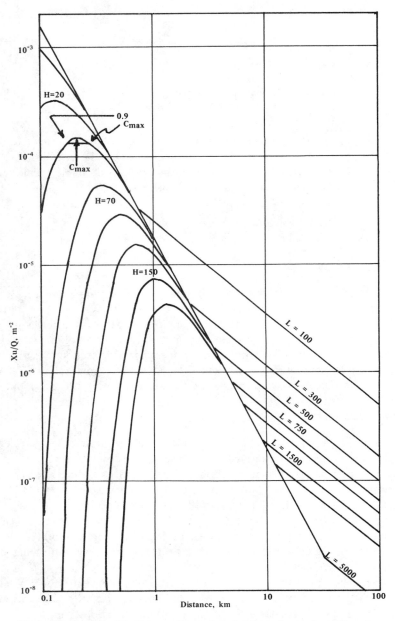

Figure 27. $\chi u/Q$ with distance for various heights of emission (H) and limits to vertical dispersion (L), B stability.[12]

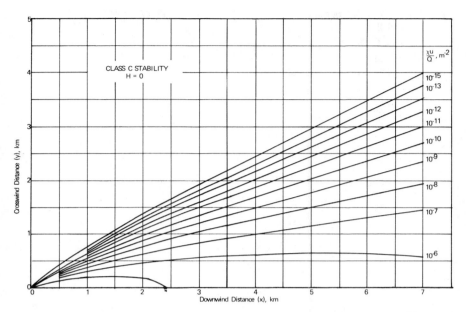

Figure 28. Isopleths of $\chi u/Q$ for a ground-level source, C stability.[12]

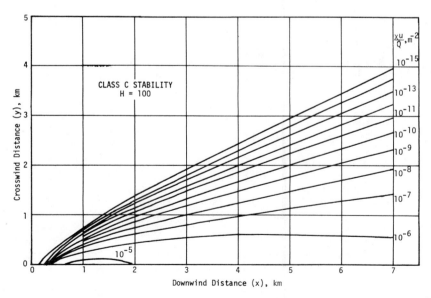

Figure 29. Isopleths of $\chi u/Q$ for a source 100 meters high, C stability.[12]

Horizontal concentration gradients in the x and y directions can be inferred from these graphs (see Turner[12] for plots for stabilities A-F.)

Another graphical solution prepared by Turner shows the location of the maximum normalized ground level concentration, $\chi u/Q$, as a function of effective height of emission, H, and stability class (see Figure 30).

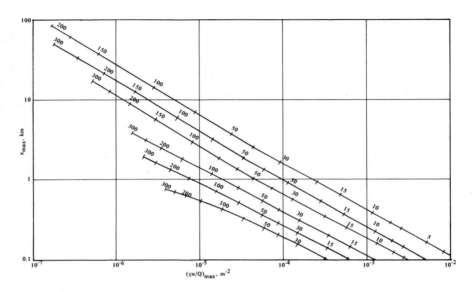

Figure 30. Distance of maximum concentration and maximum $\chi u/Q$ as a function of stability (curves) and effective height (meters) of emission (numbers).[12]

The maximum concentration can be determined by finding $\chi u/Q$ and multiplying by Q/U. According to Turner:

> In using Figure 30, the user must keep in mind that the dispersion at higher levels may differ considerably from that determined by the σ_y's and σ_z's used here. As noted, however, since σ_y generally decreases with height and U increases with height, the product of $U \sigma_y \sigma_z$ will not change appreciably. The greater the effective height, the more likely it is that the stability may not be the same from the ground to this height. With longer travel distances such as the points of maximum concentrations for stable conditions (Types E or F), the stability may change before the plume travels the entire distance.[12]

Nomographic Solutions

TVA[27,29] has presented nomographic methods of solving the Gaussian plume equations for a point source (see Figures 26, 31 and 32). These nomograms can be used to calculate the maximum plume centerline concentration for a coning plume, inversion breakup fumigation and plume trapping conditions. The following is a description of how to use the nomographs taken from the TVA reports.

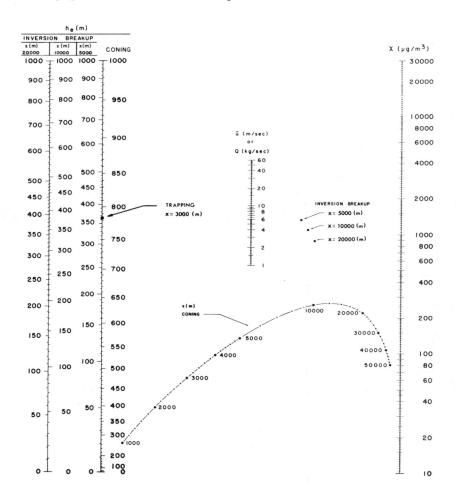

Figure 31. Plume centerline 1-hr average surface concentration, x, versus downwind distance, r, effective stack height, h_e, effluent emission rate, Q, and mean wind speed, $\bar{u}$.

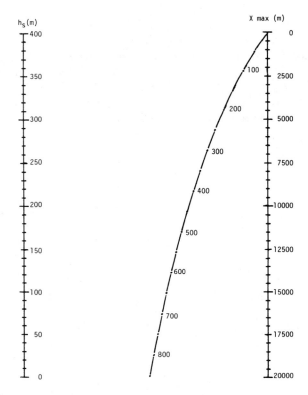

Figure 32. Distance, X_{max}, versus stack height, h_s, and h_d, for inversion breakup dispersion (from TVA).[27]

Procedure 1 — For Maximum Plume Centerline Surface Concentrations

Coning Dispersion Model:

Step 1   Project a line from the calculated G value on the left-hand G scale of Figure 26 to a value equal to h_s on the right-hand neutral Δh scale. Read the $\bar{u}$ value from the center scale. This is the $\bar{u}$ associated with the maximum plume centerline surface concentration, X_{max}. (Chi, χ, is a symbol denoting 1-hr surface sulfur dioxide concentrations in this guide.)

Step 2   Calculate effective stack height, $h_e = 2h_s$ (m)

Step 3   Locate this h_e value on the coning h_e scale of Figure 31. Project a line from this point tangent to the curved coning x scale. The point of tangency is the downwind distance where the maximum plume centerline surface concentration is expected to occur. Mark the point where the above projected line intersects the right-hand scale. From this point project a line through the $\bar{u}$ value, center scale, obtained previously from Figure 26 to the coning h_e scale. Mark this point of intersection. From this point project a line through the selected Q, center scale, to the right hand scale. Note: if Q $<$ 1 kg/sec, use Q = 1 kg/sec and the appropriate correction factor described later. Intersect value on this right-hand scale is the coning maximum 1-hr average plume centerline surface concentration, χ_{max}, in micrograms per cubic meter.

Step 4   Multiply the result by any correction factors, following the nomograms, that are applicable.

Inversion Breakup Model:

Step 1   Project a line from the calculated G value on the left-hand G scale of Figure 26 through $\bar{u}$ = 3 m/sec to the right-hand stable Δh scale. Read Δh in meters from this stable Δh scale. Calculate $h_d = \Delta h + 101$ (m).

Step 2   Locate h_s on the left-hand h_s scale of Figure 32. Locate the h_d value obtained from Step 1 on the center h_d scale. Connecting these two points, project a line to the right-hand χ_{max} scale. Intersect point is the distance downwind, χ_{max}, where the maximum inversion breakup surface concentration is expected to occur.

Step 3   Calculate effective stack height, $h_e = h_s + \Delta h$ (m).

Step 4   Locate this h_e value on the inversion breakup h_e scale of Figure 31 that represents the x (5,000, 10,000, or 20,000 m) that is most nearly equal to χ_{max}. If the x = 10,000 or 20,000 m$-h_e$ scale is utilized, project a horizontal line to the x = 5,000 m$-h_e$ scale and mark the point of intersection on the 5,000 m$-h_e$ scale. If χ_{max} is most nearly equal to x = 5,000 m, mark the h_e value obtained from Step 3 on that scale. From the resulting point on the inversion breakup x = 5,000 m$-h_e$ scale project a line through the inversion breakup point, representing the x that is also most nearly equal to χ_{max}, to the right-hand scale. Mark this point of intersection.

From this point project a line through $\bar{u}$ = 3 m/sec to the coning h_e scale. Mark this point of intersection. From this point project a line through the selected Q, center scale, to the right-hand scale. Note: If Q $<$ 1 kg/sec, use Q = 1 kg/sec and the appropriate correction factor described later. Intersect value on this right-hand scale is the inversion breakup maximum 1-hr average plume centerline surface concentration, χ_{max}, in micrograms per cubic meter.

Note: the χ_{max} is actually for x = 5,000, 10,000 or 20,000 m which may be significantly different from χ_{max}. A more accurate procedure would be to calculate two χ_{max} values, one for x $<$ χ_{max} and one for x $>$ χ_{max} and then calculate χ_{max} by linear interpolation.

Step 5   Multiply the result by any correction factors, following the nomograms, that are applicable.

Trapping Model:

Step 1   Project a line from the trapping, $X = 3,000$ m dot located on the coning h_e scale of Figure 31 through the appropriate Q, center scale, to the right-hand scale. Note: If $Q < 1$ kg/sec, use $Q = 1$ kg/sec and the appropriate correction factor described later. Intersect value on this right-hand scale is the trapping maximum 1-hr average plume centerline surface concentration, χ_{max}, in micrograms per cubic meter.

Step 2   Multiply the result by any correction factors, following the nomograms, that are applicable.

Correction Factors:

The following correction factors should be applied, if required, when using the nomograms.

Emission Rate Less than 1 kg/sec:

For emission rates less than 1 kg/sec, multiply the result from Figure 31 by actual emission rate, Q (kg/sec).

1-hr to 24-hr Averaging Time:

Multiply the result from Figure 31 by 1/3 to obtain an estimate of the 24-hr surface concentration, χ, $(\mu g/m^3)$.

Multiple-Stack Source:

Multiply result from Figure 31 by $N^{4/5}$, where N is the number of stacks of a given height.

Table 3. Correction Factor for Multiple Stack

Number of Stacks, N	Correction Factor, $N^{4/5}$
2	1.7
3	2.4
4	3.0
5	3.6
6	4.2
7	4.7
8	5.3

If a multiple stack source consists of stacks of approximately one height, and also stacks of a considerably different height, these could be treated as distinct groups of sources and the resulting concentrations from each group added to obtain the total concentration.

REFERENCES

1. Slade, D. H., Ed. *Meteorology and Atomic Energy–1968* (Oak Ridge, Tennessee: U.S. Atomic Energy Commission, Office of Information Services, 1968).
2. Bibbero, R. J. and I. G. Young. *Systems Approach to Air Pollution Control.* (New York: John Wiley & Sons, 1974).
3. Treybal, R. E. *Mass-Transfer Operations*, 2nd ed. (New York: McGraw-Hill, 1968), p. 17.
4. Hanna, S. Lecture Notes for Diffusion Meteorology, Course No. 5760, University of Tennessee (1973).
5. Smith, M. Ed. "Recommended Guide for the Projection of the Dispersion of Airborne Effluent," Amer. Soc. Mech. Eng. (1968).
6. Pasquill, F. *Atmospheric Diffusion.* (New York: Van Nostrand, 1969).
7. Smith, M. E. and I. A. Singer. "An Improved Method of Estimating Concentrations and Related Phenomena from a Point Source Emission," USAEC Report BNL-9700, Brookhaven National Laboratory (1965).
8. Briggs, G. A. "Diffusion Estimation for Small Emissions," A.T.D.L., Oak Ridge, Tennessee, draft report (May 1973).
9. Tennessee Valley Authority. "Summary of Tennessee Valley Authority Atmospheric Dispersion Modeling," presented by the Conference on the TVA Experience of International Institute for Applied Systems Analysis, Schloss Laxenburg, Austria (October 1974).
10. Hilst, G. R. and C. L. Simpson. "Observations of Vertical Diffusion Rates in Stable Atmospheres," *J. Meteorol.* 15(1):125-126.
11. Islitzer, D. In *Recommended Guide for the Production of the Dispersion of Airborne Effluent*, M. Smith, Ed. (Amer. Soc. Mech. Eng., 1968).
12. Turner, D. "Workbook of Atmospheric Dispersion Estimates," U.S. EPA Publication No. AP-26 (1970).
13. Sutton, O. G. "A Theory of Eddy Diffusion in the Atmosphere," *Proc. Roy. Soc. London* 135:143-165 (1932).
14. Beaton, J. L., A. J. Ranzieri, E. C. Shirley and J. B. Skog. *Air Quality Manual–Vol. IV, Mathematical Approach to Estimating Highway Impact of Air Quality*, (Sacramento, California: California Department of Public Works, Division of Highways, April 1972).
15. U.S. Environmental Protection Agency. "A User's Guide for Hiway," (Research Triangle Park, North Carolina: National Environmental Research Center, Meteorology Lab, November 1974).
16. Gilbert, J. and W. T. Davis. "A Comparison of EPA and California Line Source Diffusion Models," unpublished paper, Department of Environmental Engineering, University of Tennessee (1974).
17. Miller, M. E. and G. C. Holzworth. "An Atmospheric Diffusion Model for Metropolitan Areas," *J. Amer. Poll. Control Assoc.* 17(1):46-50 (1967).
18. Calder, K. L. "A Climatological Model for Multiple Source Urban Air Pollution," *Proc. 2nd Meeting of the Expert Panel on Air*

Pollution Modeling, NATO Committee on the Challenges of Modern Society, Paris, France (July 26-27, 1971).

19.   Gifford, F. A. and S. R. Hanna.   "Modeling Urban Air Pollution," *Atmos. Envir.* 7:131-136 (1973).

20.   Mancuso, R. L. and F. L. Ludwig.   "User's Manual for the APRAC-1A Urban Diffusion Model Computer Program,"   Contract No. CAPA-3-68 (1-69), Stanford Research Institute, Menlo Park, California, Project No. 8563 (September 1972).

21.   Briggs, G. A.   "Plume Rise,"   U.S. Atomic Energy Commission Division of Information (1969).

22.   Fay, J. A. *et al.*   "A Correlation of Field Observations of Plume Rise," *J. Air Pollution Control Assoc.* 20:391-397 (1970).

23.   Hoult, D. P., J. A. Fay and L. J. Forney.   "A Theory of Plume Rise Compared with Field Observations," *J. Air Pollution Control Assoc.* 19:585-590 (1969).

24.   Rupp, A. F., S. E. Beall, L. P. Bornwasser and D. F. Johnson.   "Dilution of Stack Gases in Cross Winds,"   U.S. AEC Report AECD-1811 (CE-1620), Clinton Laboratories (1948).

25.   Briggs, G. A.   "Some Recent Analyses of Plume Rise Observations," *Proc. 2nd Internat. Clean Air Conf.*, Washington, D.C.   (New York: Academic Press, 1970).

26.   Briggs, G. A.   Class Notes from Course Taught at University of Tennessee, Knoxville (May 1975).

27.   Montgomery, T. L. *et al.*   "A Simplified Technique Used to Evaluate Atmospheric Dispersion of Emissions from Large Power Plants,"   In: *Power Generation—Air Pollution Monitoring and Control*, K. E. Noll and Wayne T. Davis, Eds.   (Ann Arbor, Michigan: Ann Arbor Science Publishers, Inc., 1976).

28.   Montgomery, T. L. *et al.*   "Results of Recent TVA Investigations of Plume Rise," *J. Air. Pollution Control Assoc.* 22(10) (October 1972).

29.   *Engineering Guide for Evaluating Dispersion of Sulfur Dioxide Emissions.*   Internal document of the Tennessee Valley Authority (1971).

CHAPTER III

GENERAL PROCEDURES FOR
AIR MONITORING SURVEY DESIGN

INTRODUCTION

Air quality surveys involve air quality monitoring, meteorological monitoring, calibration and data acquisition systems. A well-designed air quality survey will address the following items in the order listed:

1. Set the objectives of the air monitoring investigation.
2. Determine the physical parameters to be measured.
3. Set the network specifications including the location of air monitoring stations, the duration of the study and sampling schedules, and the air sampling method to be used.
4. Set the specifications for the individual stations in the network including the equipment needed to conduct the study, the method and frequency for equipment calibration and the data recording methods.
5. Determine the type of data analysis to be performed and method of data reporting.

It should be recognized that these steps are interdependent and that a properly designed survey would consider each step in terms of the effect on the other parts of the survey design.

AIR MONITORING NETWORK SPECIFICATIONS

The sum total of all air monitoring stations, meteorological stations, calibration equipment and data acquisition equipment required to meet the total objective of an air quality survey represent the air monitoring network.

In order to understand the interrelationships between all of the component parts of the network and allow decisions to be made about the number and type of each piece of equipment and the interdependence of

51

the equipment in meeting the survey objectives, a set of specifications for the network must be developed early in the planning process.

Air monitoring network specifications should include the number of sites to be monitored, air pollution and meteorological measurements required at each site, the duration of the study, and manpower requirements. Network specifications should be determined in light of known limitations of physical, engineering, economic and human factors as well as limitations due to completion deadlines or time delays in equipment procurement. Time, manpower and budget constraints may significantly affect the feasibility of utilizing specific techniques or equipment to conduct the study.

Considering all of the stated sampling requirements and available resources, the types of air quality and meteorological monitoring equipment to be employed can be identified, including the total number of samplers and analyzers for each type of measurement to be performed. To accomplish this, available sampling methods should be evaluated. For example, the number of continuous analyzers versus semi-automatic samplers versus manual sampling methods should be investigated. Obviously, the needed mobility of air sampling equipment should be considered and such factors as the high capital costs of continuous analyzers versus the high operating costs of semi-automatic or manual methods should be optimized for cost effectiveness, where possible.

The calibration system to be used to insure accurate data from all the analyzers should be considered next. Each station can be equipped with calibration hardware. Portable instruments can be calibrated at a central laboratory, or permanent instruments can be calibrated with a portable calibration device. The choice of needed calibration equipment will depend on the number of stations, the distance separating the stations and the frequency at which calibrations are performed. A typical integrated calibration system might consist of sophisticated dynamic calibration apparatus located at a central laboratory for performing periodic primary calibrations, with several portable calibrators or span gas bottles used for frequent on-site single-point calibration checks.

Once the air monitoring and calibration specifications have been fixed, the requirements of the data acquisition system can be evaluated. The data rates and data quantities generated by the monitoring network should be determined as a whole in order to calculate the data handling capacity needed for the overall sampling network. Data recording equipment, which is ideal for a single monitoring station, may be too expensive for use at numerous monitoring stations. Frequently it is less expensive to replace a variety of special devices with a central general purpose control element, even though the speed and flexibility requirements of the system do not require it. Data telemetry and central data processing systems may represent the most efficient approach for monitoring networks with many remote sampling stations employing continuous air pollution analyzers. Conversely, for single station applications, strip chart recorders with manual data reduction may represent the most cost effective data collection method.

At this point, air monitoring *station specifications* can be developed for each monitoring site in the network. Station specifications include the setting of sampling objectives for each station and the selecting of compatible hardware components for each station, resulting in an integrated air monitoring sampling station design.

There are nine discrete component elements of air monitoring station design: the sampling probe and tubing, sample preconditioning, sample collectors, continuous air pollution analyzers, calibration equipment, flow control and flow meters, air movers, data acquisition equipment and equipment shelters (see Figure 33).

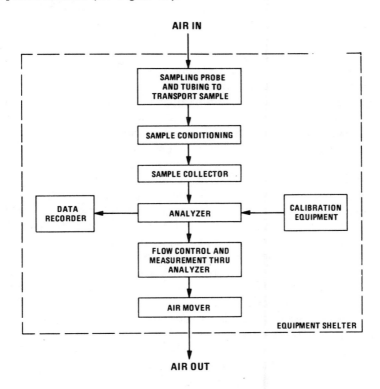

Figure 33. Hardware components of an air quality monitoring station.

The sequence of selection of station equipment should be arranged so that primary components are chosen before supportive or supplementary hardware. For example, equipment shelters should be designed after the total amount of equipment to be housed in the shelter has been determined; thus, the shelter will be properly sized, with adequate electrical wiring, plumbing, instrument mounting and storage space as required by hardware components.

The first step in hardware selection is to choose air samplers and air pollution analyzers. Similar instruments can be compared on a cost basis, or by using manufacturers' instrument operation specifications. Instrument selection can be assisted by consulting the technical literature for reports on instrument performance tests or by obtaining personal recommendations from experts in the air monitoring instrumentation field. An extensive compilation of air monitoring instrument hardware specifications has been prepared by Lawrence Berkeley Laboratory,[1] which may be especially useful for selecting instrument hardware. Also, the EPA has published *Unacceptable Methods of Measurement for Criteria Pollutants*,[2] which should be consulted before selecting analyzers.

Second, calibration equipment should be selected that is compatible with the air monitoring instruments. The critical criterion is that calibration equipment must produce pollutant gas concentrations at desired levels and in sufficiently large quantities to calibrate the monitoring equipment. Problems can arise if the monitoring instrument employs a high sample flow rate while the calibration device generates a low flow rate. Another common problem occurs when the pollutant gas concentration from the calibrator is too high to calibrate the monitoring instrument in the desired operating range.

Third, data acquisition equipment should be compatible with the air monitoring hardware previously selected. Voltage output ranges from monitoring equipment must match the voltage input ranges of the data acquisition equipment. The data acquisition system must be properly sized in order to accommodate the total data output from all air monitoring equipment. When monitoring stations are operated over long periods unattended, the data system must provide enough storage capacity to handle all the output data generated by the sensor. Data acquisition systems should also provide methods by which instrument calibration and maintenance information can be stored along with data from the air pollution and meteorological sensors.

Fourth, sample preconditioning devices should be selected according to the requirements of the air pollution analyzers employed. The installation of sample preconditioning equipment in the sampling lines may cause difficulties if the plumbing requirements are not anticipated before designing

the sampling probe hardware. Since sample preconditioner plumbing requirements may affect the sampling probe or manifold design, the selection
of preconditioning equipment should precede the sample handling system
design.

Fifth, the entire sample handling system, which consists of the sampling probe and tubing, flow meters, flow controllers and air mover equipment, should be designed as an integral unit with compatible components.
The selection of the sampling probe and the sample tubing depends on
the type of sampling to be conducted, the type of air pollutants to be
measured and the sample volume flow rates required by air samplers
operated within the station. The important design criteria is that the
tubing be sized to handle the air sample flow rates required by the instruments and that the materials of construction not affect the air sample
before analysis. Flow meters should be employed on all sampling lines
where the flow rate requires periodic monitoring. Flow meters should be
chosen according to both the flow rate to be measured and the materials
of construction of the meter. Flow controlling devices and air movers
should be properly matched to provide a continuous regulated sample
flow rate through all "legs" of the sample handling system.

Sixth, the size of shelter needed depends on the total amount of
equipment to be housed. An inventory of all equipment required, such
as space, electrical power needs, environmental control requirements,
plumbing needs and vibration effects, can be used to insure an adequate
shelter size, with adequate electrical wiring, plumbing, instrument mounts
and storage space as needed. Station mobility requirements must also be
considered in light of the quantity of monitoring to be conducted and
the supportive equipment required for remote mobile station monitoring.

REVIEW OF CURRENT AIR MONITORING
SURVEY DESIGN PRACTICES

It has been generally recognized that the choice of monitoring site depends on the objective of the monitoring to be performed. EPA[3]
recognizes the following as principal objectives for monitoring:

1. To judge compliance with and/or progress made toward meeting
   ambient air quality standards.
2. To activate emergency control procedures to prevent air pollution
   episodes.
3. To observe pollution trends throughout the region including the
   nonurban areas. (Information on the nonurban areas is needed to
   evaluate whether air quality in the cleaner portions of a region is
   deteriorating significantly and to gain knowledge about background
   levels.)

4. To provide a data base for application in evaluation of effects; urban, land use and transportation planning; development and evaluation of abatement strategies; and development and validation of diffusion models.

In order to accomplish these objectives, criteria for air monitoring network sites and individual monitoring stations are required.

Station Siting Criteria

Most of the criteria that have been published regarding the location of an individual sampler or a continuous monitoring station are aimed at insuring a representative sample without undue influence from the immediate surroundings. Again quoting from *AP-98*[3] :

> No definitive information is available concerning how air quality measurements are affected by the nearness of buildings, height from ground and the like. There are, however, general guidelines that should be considered in site selection.
>
> 1. Uniformity in height above ground level is desirable for the entire network within the region. Some exceptions may include canyons, high-rise apartments and sites for special-purpose samplers.
>
> 2. Constraints to airflow from any direction should be avoided by placing inlet probes at least three meters from buildings or other obstructions. Inlet probes should be placed to avoid influence of convection currents.
>
> 3. The surrounding area should be free from stacks, chimneys or other local emission points.
>
> 4. An elevation of three to six meters is suggested as the most suitable for representative sampling, especially in residential areas. Placement above three meters prevents most reintrainment of particulates, as well as the direct influence of automobile exhaust.

Pooler,[4] in reference to sites chosen for the St. Louis Regional Air Pollution Study (RAPS), the objective of which is the development and validation of regional air quality simulation models, emphasizes the need for a representative site. "Ideally, sites chosen for the RAPS program should each be representative of a 1 km² area of downtown St. Louis." However, because of the inhomogeneities of emissions throughout the study area, Pooler has restated his criteria to "avoidance of domination of a site by one or more local sources." He recommends that a site should be 1 km or more from a major traffic artery, it should not be in the lee of a building where downwash of effluents from that building or its immediate neighbors might occur, it should not be located within a local topographic depression, and states that busy parking lots are poor sampling locations.

Most authors recognize the need for accessibility and availability of utilities for each site, which often eliminates many potential locations. Based on his experience with vandalism and theft at Chicago monitoring stations, Harrison[5] places the highest priority on security, stating: "Experience in urban areas leads one quickly to the unfortunate realization that the first criterion for placement for a sensor for any purpose is security."

The Highway Research Board[6] has published EPA recommendations, which are more definitive in describing acceptable sampling locations. Table 4 lists specific guidelines for locating air monitoring instruments in areas of estimated maximum pollutant concentration as given by EPA:[6]

> Sampling station guidelines are different for defining average CO concentration for one hour and eight hours because people would not ordinarily be exposed to CO concentrations that occur in a downtown area having high traffic density for a period of eight hours. When only a single sampling site is established to satisfy the minimum air quality surveillance requirement of the implementation plan, a site should be chosen that meets the guidelines for eight hour averaging time. Distance from the street is specified in the sampling location guidelines for CO because of the strong dependence on nearness to the street and CO concentration. For the same reason, height from the ground of the air inlet is more restrictive than for the other pollutants. It is desirable, however, to sample as close as possible to the breathing zone within practical considerations; sampling height limitations are specified accordingly for those pollutants. There are no well established meteorological dispersion models currently available for selecting areas of expected maximum concentration for the secondary pollutants. Selection of high concentration areas described in Table 4, for those pollutants is based on available information on reaction kinetics of atmospheric photochemical reactions involving hydrocarbons, nitrogen oxides, and oxidants; atmospheric data on diurnal variation in pollutant concentration; distribution of primary mobile sources of pollution; and meteorological factors. A minimum distance away from major traffic arteries and parking areas is specified for the oxidant monitoring site because NO emissions from motor vehicles consume atmospheric ozone. NO_2 is considered to be both a primary-stationary-source pollutant and a secondary pollutant, and air monitoring stations for this pollutant should be located consistent with the respective station location guidelines. Differences in horizontal and vertical clearance distance are based on increased probability of reaction between reactive gases and vertical surfaces.

The need for nationally standardized criteria for locating monitoring stations has been suggested in two papers by Ott et al.[7,8] His study of carbon monoxide concentrations in San Jose[7] indicates that air monitoring stations located near city streets "give drastically different results when moved only a block or two." This raises serious implications regarding the representativeness of these measurements to air quality in a large physical

Table 4. Sampling Location Guidelines for Areas of Estimated Maximum Pollutant Concentration[6]

Pollutant Category	Pollutant	Station Location	Position of Air Inlet		
			Height[a] (ft)	Vertical[b] (ft)	Horizontal[c] (ft)
Primary stationary source	SO_2 NO_2 Particulate matter	Determined from atmospheric diffusion model, historical data, emission density, or other information and representative of population exposure	<50	>3	>5
Primary mobile source	CO[d]	Representing area having high traffic density, slowly moving traffic, obstructions to air flow (tall buildings), and pedestrian population such as major downtown traffic intersection (<20 ft from street curb)	<15	>3	>3
	CO[e]	Representing area having high traffic density in residential area such as major thorough-fare in center city or suburban area (<50 ft from street curb)	<15	>3	>3
Secondary	O_3	Representing residential area downwind of downtown area (5 to 15 miles from downtown and > 300 ft from major traffic arteries or parking areas)[f]	<50	>3	>5
	NO_2	Representing residential area downwind of downtown area (<5 miles from downtown)[f]	<50	>3	>5

[a]From ground.
[b]Clearance above supporting structure
[c]Clearance beyond supporting structure; not applicable where air inlet is located above supporting structure.

[d]1-hr averaging time
[e]8-hr averaging time
[f]Downwind of prevailing daytime wind direction during oxidant season.

area surrounding the station. Because of the complex nature of the spatial variations in urban CO concentrations, Ott[8] suggests a "dual monitoring system" consisting of a "background exposure station" located outside the range of influence of nearby traffic (greater than 100 meters from the curb) and a "pedestrial exposure station" with a sampling probe above the sidewalk, 0.5 meters away (horizontally) from the curb and 3 ± 0.5 meters above the sidewalk. In the dual monitoring approach, the pedestrian exposure station provides a measure of the exposure of individuals on downtown streets. The background exposure station measures concentrations occurring over a large physical area of the city.

Ott goes on to say that few of today's air monitoring sites could be classified into either of the above two categories. He states: "My examination of many existing sites reveals that too many fall between categories—for example, too far from the curb to give an estimate of pedestrian exposure and too close to the street to avoid serious influence from traffic. In my view, such sites give ambiguous measurements. . . . It may not be possible to determine if an urban area is or is not in compliance with an air quality standard, as presently defined, using data from existing air monitoring stations." As a solution he recommends adoption of six standard classifications of criteria for siting monitoring stations (see Table 5).

Network Siting Criteria

In general, available information in the literature recognizes that the design of an air monitoring network involves a trade-off between what is considered desirable from a strictly technical point of view and what is feasible with the available resources. An ideal network will usually require more resources than are available. The objective is to design the least cost monitoring network still capable of meeting the major surveillance requirements. Network design should attempt to maximize the effectiveness of a minimally adequate network by careful selection of sampling sites, sampling schedules and use of both semi-automatic and automatic (continuous) monitoring instruments.

According to EPA,[3]

> Knowledge of the existing pollution levels and patterns within the region is essential in network design. The areas of highest pollution levels must be defined, together with geographical and temporal variations in the ambient levels. Isopleth maps of ambient concentrations derived from past sampling efforts and/or from diffusion modeling are the best tools for determining the number of stations needed and for suggesting the station locations. Additionally, information on meteorology, topography, population distribution, present and projected land uses, and pollution

Table 5. Recommended Criteria for Siting Monitoring Stations[8]

Station Type	Description
Type A	**Downtown Pedestrian Exposure Station.** Locate station in the central business district (CBD) of the urban area on a congested, downtown street surrounded by buildings (*i.e.*, a "canyon" type street) and having many pedestrians. Average daily travel (ADT) on the street must exceed 10,000 vehicles/day, with average speeds less than 15 miles/hr. Monitoring probe is to be located 1/2 m from the curb at a height of $3 \pm 1/2$ m.
Type B	**Downtown Background Exposure Station.** Locate station in the central business district (CBD) of the urban area but not close to any major streets. Specifically, no street with average daily travel (ADT) exceeding 500 vehicles/day can be less than 100 m from the monitoring station. Typical locations are parks, malls or landscaped areas having no traffic. Probe height is to be $3 \pm 1/2$ m above the ground surface.
Type C	**Residential Population Exposure Station.** Locate station in the midst of a residential or suburban area but not in the central business district (CBD). Station must not be less than 100 m from any street having a traffic volume in excess of 500 vehicles/ day. Station probe height must be $3 \pm 1/2$ m.
Type D	**Mesoscale Meteorological Station.** Locate station in the urban area at appropriate height to gather meteorological data and air quality data at upper elevations. The purpose of this station is not to monitor human exposure but to gather trend data and meteorological data at various heights. Typical locations are tall buildings and broadcasting towers. The height of the probe, along with the nature of the station location, must be carefully specified along with the data.
Type E	**Nonurban Background Station.** Locate station in a remote, non-urban area having no traffic and no industrial activity. The purpose of this station is to monitor for trend analyses, for nondegradation assessments and for large-scale geographical surveys. The location or height must not be changed during the period over which the trend is examined. The height of the probe must be specified along with the data. A suitable height is $3 \pm 1/2$ m.
Type F	**Specialized Source Survey Station.** Locate station very near a particular air pollution source under scrutiny. The purpose of the station is to determine the impact on air quality, at specified locations, of a particular emission source of interest. Station probe height should be $3 \pm 1/2$ m unless special considerations of the survey require a nonuniform height.

sources is extremely helpful in network design. . . . Information on emission densities and/or land use can be used together with wind-rose data to pinpoint areas of expected higher concentrations. Topographical maps provide additional information on wind flow and pollution dispersion characteristics. Maps of population distribution are essential in locating key stations for monitoring during episodes.

The following are guidelines suggested as important network site selection criteria:

1. The priority area is the zone of highest pollutant concentration within the region. One or more stations are to be located in this area.

2. Close attention should be given to densely populated areas within the region, especially when they are in the vicinity of heavy pollution.

3. For assessing the quality of air entering the region, stations must also be situated on the periphery of the region. Meteorological factors such as frequencies of wind direction are of primary importance in locating these stations.

4. For determining the effects of future development on the environment, sampling should be undertaken in areas of projected growth.

5. A major objective of surveillance is evaluation of the progress made in attaining the desired air quality. For this purpose, sampling stations should be strategically situated to facilitate evaluation of the implemented control tactics.

6. Some information of air quality should be available to represent all portions of the region.

The air quality surveillance network should consist of stations that are situated primarily to document the highest pollution levels in the region, to measure population exposure, to measure the pollution generated by specific classes of sources, and to record the nonurban levels of pollution. Many stations will be capable of meeting more than one of these criteria. . . . Common sites, although not necessarily ideal for each pollutant, may be selected to provide adequate coverage for the pollutants of concern. Each pollutant, however, should be considered individually during the design phase to pinpoint pockets of high pollution that otherwise might be overlooked.

Pooler[4] has described the criteria used to select the air monitoring network for the Regional Air Pollution Study (RAPS) in St. Louis. According to Pooler, sites chosen were based on the "conflicting requirement of extended regional coverage versus close spacing of sites to achieve desired spatial resolution." The basic objective of RAPS is mesoscale model validation in a complex urban area. Pooler states:

This objective leads directly to the requirement that there be generated an extensive base of air quality measurements against which calculated values may be compared. Extensive spatial coverage is necessary to include the entire metropolitan area, and simultaneous measurements should be made at all sites for a long enough interval of time to include all the varying large-scale weather situations that might reasonably be expected to occur. Measurements should be continuous, but with a time resolution short enough to be compatible with the resolution available from models. . . .

Since the network is for the purpose of model development, the resources to be put into the network must be balanced against resources required for other components in a total modeling effort, namely, the requirements for emission data and the requirements for studies of the atmospheric processes of dispersion and transformations. . . .

The relationship between input resources and the value of data to be obtained is generally nonlinear, and thus there exist no simple formulations by which an optimum mix of fixed network measurements, special measurements, analytical studies and the duration of each measurement and study may be derived. However, there are minima, which, if not met result in substantial losses of value in the output. One minimum we agreed on was the number of sites for such a network. We were originally thinking in terms of 40 to 50 sites, but when we began estimating costs, we were forced to consider what minimum number could serve our purposes, and arrived at the number 25. Our reasoning was quite subjective, but can be demonstrated by trying to place pins on a map, subject to the general criteria to be discussed below. . . .

Given the magnitude of the network and the measurements to be made, a schematic network layout can be devised, subject to several general criteria. First, if we are to model the influence of emissions from an urban area on air quality, we should know what the concentrations are in the air approaching the area, i.e., at least one station should be in an upwind sector regardless of the wind direction. Since upwind-downwind comparisons can be made of air quality as well as radiation parameters, each upwind site should have a downwind counterpart. This criterion can be met by having four rural sites at approximately 90° azimuth spacing, with the alignments along the most frequent wind directions, namely, west-northwest to east-southeast, and south to north.

The second general criterion is that the site spacing should depend on both concentrations and their spatial gradients, with the minimum spacing where the concentrations and gradients are highest. . . . Calculations of annual SO_2 and particulate patterns indicate the highest concentrations and gradients occur generally within about 10 km of the Arch, and, therefore, for these pollutants the network should be concentrated within this area. For primary pollutants from mobile sources, the patterns should more closely resemble patterns of vehicle emissions with a maximum in the downtown area, but generally more spread out than the calculated SO_2 and particulate patterns. Secondary mobile source pollutants should show more seasonal influence, since the higher solar intensities and air temperatures in summer will aid in their formation. We guess at this time that the maximum of O_3 will probably occur beyond the most distant sites, since it takes a number of hours for the O_3 maximum to develop. There should be occasions when the NO_2 maximum occurs within the bounds of the network, and, considering the summer wind patterns, should most frequently be found.

Taking these various factors into account, a general station siting plan consisting of a central station surrounded by loosely arrayed rings of stations at increasing radial spacings appears to be a suitable arrangement. . . . The center of this array should be somewhat north of the downtown area of St. Louis, with the closest spacing of sites being about 4 km. The spacing should increase to 10 to 15 km at the outermost sites, excluding the four rural sites.

An air monitoring network employed by the Tennessee Valley Authority (TVA) for evaluating the maximum SO_2 concentration occurring downwind of the Paradise Steam Plant has been described by Montgomery *et al.*[9] This network employed 14 digital recording SO_2 instruments located within the 22.5-degree prevailing downwind sector within 17 km of the plant. Instruments were located on radial arcs, with the largest number in the expected area of maximum ground level SO_2 concentrations (*i.e.*, within 4 and 5 km). There were also several other SO_2 instruments located in the peripheral areas around the steam plant (see Figure 34).

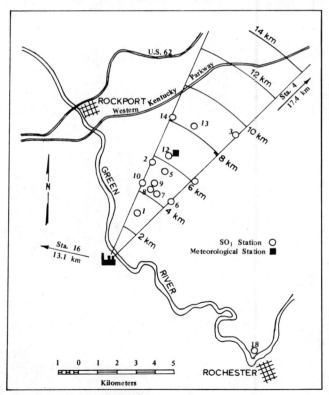

Figure 34. Paradise steam plant air monitoring network.

Table 6. Air Quality Surveillance System Requirements[10]

Classification of Region	Pollutant	Measurement Method[j]	Minimum Frequency of Sampling	Region Population	Minimum Number of Air Quality Monitoring Sites[h]
I	Suspended particulates	High volume sampler	One 24-hr sample every 6 days[a]	Less than 100,000	4+0.6 per 100,000 population[c]
				100,000-1,000,000	7.5+0.25 per 100,000 population[c]
				1,000,001-5,000,000	12+0.16 per 100,000 population[c]
				Above 5,000,000	One per 250,000 population[c] up to 8 sites
		Tape sampler	One sample every 2 hr		4
	Sulfur dioxide	Pararosaniline or equivalent[d]	One 24-hr sample every 6 days (gas bubbler)[a]	Less than 100,000	3
				100,000-1,000,000	2.5+0.5 per 100,000 population[c]
				1,000,001-5,000,000	6+0.15 per 100,000 population[c]
				Above 5,000,000	11+0.05 per 100,000 population[c]
			Continuous	Less than 100,000	1
				100,000-5,000,000	1+0.15 per 100,000 population[c]
				Above 5,000,000	6+0.05 per 100,000 population[c]
	Carbon monoxide	Nondispersive infrared or equivalent[e]	Continuous	Less than 100,000	1
				100,000-5,000,000	1+1.15 per 100,000 population[c]
				Above 5,000,000	6+0.05 per 100,000 population[c]
	Photochemical oxidants	Gas phase chemiluminescence or equivalent[f]	Continuous	Less than 100,000	1
				100,000-5,000,000	1+0.15 per 100,000 population[c]
				Above 5,000,000	6+0.05 per 100,000 population[c]
	Nitrogen dioxide	24-hr sampling method (Jacobs-Hochheiser method)	One 24-hr sample every 14 days (gas bubbler)[b]	Less than 100,000	3
				100,000-1,000,000	4+0.6 per 100,000 population[c]
				Above 1,000,000	10
II	Suspended particulates	High volume sampler	One 24-hr sample every 6 days		3
		Tape sampler	One sample every 2 hr		1
	Sulfur dioxide	Pararosaniline or equivalent[d]	One 24-hr sample every 6 days (gas bubbler)[a]		3
			Continuous		1

Table 6, Continued

Classifi-cation of Region	Pollutant	Measurement Method	Minimum Frequency of Sampling	Region Population	Minimum Number of Air Quality Monitoring Sites[h]
III[g]	Suspended particulates	High volume sampler	One 24-hr sample every 6 days[a]		1
	Sulfur dioxide	Pararosaniline or equivalent[d]	One 24-hr sample every 6 days (gas bubbler)[a]		1

[a] Equivalent to 61 random samples per year.

[b] Equivalent to 26 random samples per year

[c] Total population of a region. When required number of samplers includes a fraction, round-off to nearest whole number.

[d] Equivalent methods are (a) Gas Chromatographic Separation-Flame Photometric Detection (provided Teflon is used throughout the instrument system in parts exposed to the air stream), (2) Flame Photometric Detection (provided interfering sulfur compounds present in significant quantities are removed), (3) Coulometric Detection (provided oxidizing and reducing interferences such as O_2, NO_2, and H_2S are removed), and (4) the automated Pararosaniline Procedure.

[e] Equivalent method is Gas Chromatographic Separation-Catalytic Conversion-Flame Ionization Detection.

[f] Equivalent methods are (1) Potassium Iodide Colorimetric Detection (provided a correction is made for SO_2 and NO_2), (2) UV Photometric Detection of Ozone (provided compensation is made for interfering substances), and (3) Chemiluminescence Methods differing from that of the reference method.

[g] It is assumed that the federal motor vehicle emission standards will achieve and maintain the national standards for carbon monoxide, nitrogen dioxide and photochemical oxidants; therefore, no monitoring sites are required for these pollutants.

[h] In interstate regions, the number of sites required should be prorated to each state on a population basis.

[i] All measurement methods, except the Tape Sampler method, are described in the national primary and secondary ambient air quality standards published in the *Federal Register* on April 30, 1971 (36 F.R. 8186). Other methods together with those specified under footnotes (d), (e) and (f) will be considered equivalent if they meet the performance specifications.

This monitoring network was used to determine the types of meteorological conditions that produced the highest ground level SO_2 concentrations and the distances downwind of the plant where they occurred.

Regarding the number of stations required by a network, rules and regulations published in the *Federal Register*[10] state implementation plans must include provisions for surveillance of ambient air quality. These surveillance requirements include the minimum number of stations required to monitor criteria pollutants as a function of the region population. These minimum requirements are reproduced in Table 6.

REFERENCES

1. "Instrumentation for Environmental Monitoring—Air," Lawrence Berkeley Laboratory, University of California, Berkeley (December 1973).
2. U.S. Environmental Protection Agency. "Designation of Unacceptable Analytical Methods of Measurements for Criteria Pollutants," EPA Publication No. OAQPS 1.2-018 (May 1974).
3. "Guidelines: A/Q Surveillance Networks," AP-98 (May 1971).
4. Pooler, F. "Network Requirements for the St. Louis Regional Air Pollution Study," *J. Air Pollution Control Assoc.* (March 1974).
5. Harrison, P. R. "Considerations for Siting A/Q Monitors in Urban Areas," presented at 65th APCA Meeting (June 1972).
6. "Highways and Air Quality," Special Report 141, Highway Research Board (1973).
7. Ott, W. and R. Eliassen. "A Survey Technique for Determining the Representativeness of Urban Air Monitoring Stations with Respect to Carbon Monoxide," *J. Air Pollution Control Assoc.* (August 1973).
8. Ott, W. R. "Development of Criteria for Siting Air Monitoring Stations," presented at 68th APCA Meeting (June 1975).
9. Montgomery, T. L. *et al.* "Controlling Ambient SO_2," *J. Metals* (June 1973).
10. *Federal Register.* "Implementation Plans," 36FR15486 .

CHAPTER IV

AIR POLLUTION REGIMES

In analyzing the spatial distribution of pollution concentrations it is important to recognize three separate air pollution regimes: the microscale regime, the mesoscale regime and the macroscale regime. The regimes can be defined quantitatively according to the magnitudes of the horizontal pollution concentration gradient. To fully understand the nature of each regime both definitions are combined.

1. **Microscale:** The microscale air pollution regime is any air mass exhibiting ground level air pollution concentrations that deviate by greater than 20% over linear distances up to 100 m. The microscale air pollution regime represents a relatively small air mass exhibiting large variations in ground level air pollution concentrations. This phenomenon usually occurs very near sources of air pollution when the rate of increasing atmospheric dispersion with downwind distance is very great.

2. **Mesoscale:** The mesoscale air pollution regime is any air mass exhibiting ground level air pollution concentrations that deviate by less than 20% over linear distances between 100 m and 10,000 m. The mesoscale regime represents a "community size" air mass exhibiting fairly homogeneous ground level air pollution concentrations—for example, "local background" ambient air pollution concentrations that result, especially within urban areas, from the emission of relatively small quantities of air pollutants from a large number of ground level sources (*i.e.,* automobiles, residential and commercial space heating furnaces, and even numerous small industrial sources). These "local background" concentrations can vary considerably at different locations within an urban area. Concentration gradients greater than 20% per 100 m indicate the presence of a microscale air pollution regime superimposed upon the mesoscale regime.

3. **Macroscale:** The macroscale air pollution regime is any air mass exhibiting ground level air pollution concentrations that deviate by less than 20% over linear distances greater than 10,000 m. An

67

example of a macroscale regime is "regional background" air pollution concentrations, which can be fairly homogeneous over linear distances from tens to hundreds of kilometers. This does not mean, however, that there are not locations within the macroscale regime where ambient air pollution concentrations deviate by more than 20% from regional background levels. These deviations can and do occur when mesoscale and microscale air pollution regimes are superimposed upon the regional background air pollution concentration regime

The combined effects of the three pollution regimes on ambient pollution levels are illustrated in Figure 35 for highway sources of carbon monoxide (CO). Shown in the figure are the location of local streets, arterials and major high-capacity freeway routes for a hypothetical urban area. On the right side of the figure is a plot of the CO concentrations that might be observed at ground level for winds blowing "across town" as illustrated. Initially, the air mass contains CO at only the regional background concentrations, but as the air mass passes over numerous city streets where automobiles are being operated, the CO levels increase. In the case of city streets where emission sources are numerous, but diffuse, the horizontal distribution of concentrations will reflect a rather smooth profile of slightly increasing concentrations at greater distance from the upwind edge of the urban area source. This smooth horizontal concentration profile represents changes in pollution levels within the mesoscale regime, and is defined as the local background concentration.

At *major arterial streets*, with high traffic volumes, a peak in the concentrations above the background level will occur. The influence of this line source is significant within a horizontal distance of roughly 100 m from the edge of the highway. Throughout this distance, atmospheric dispersion of the pollutants occurs. At distances beyond 100 m only a small additional contribution to the local background concentration is observed.

A similar but greater air pollution concentration anomaly occurs at the location of a high-capacity freeway due to the higher pollutant emission rate. As in the case of the arterial street, the ground level pollution concentration returns to near local background downwind from the freeway. This geographical area near major streets and freeways where the ambient pollution levels are above the local background concentrations defines the microscale regime of the pollution concentration distribution. This same elevation in concentrations occurs near large point sources of pollution.

Pollution levels in the proximity of major sources consist of three component parts, the microscale concentration (contributed directly by the nearby source), the mesoscale concentration (the local background concentration due to area wide sources of pollutant emissions), and the

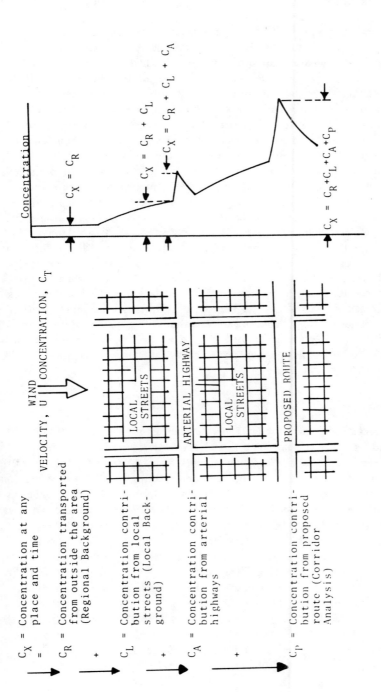

Figure 35. Background air quality definition.

macroscale concentration (the regional background concentration due to distant pollutant sources). Frequently the macroscale pollution concentration is so low that this term can be ignored, leaving only two components to any ground level observation—the local background air pollution concentration plus the microscale concentration due to a nearby source. Air monitoring sites located outside the microscale regime measure only mesoscale and macroscale concentrations (background). Sampling sites located within the microscale regime measure the concentration due to the combined microscale, mesoscale, and macroscale regimes.

Since concentrations at one site contain several component parts and because they can vary considerably at different locations within the microscale and the mesoscale regimes, it is important that monitoring sites be carefully selected. The selection of a site requires consideration of the regimes being sampled. This is important for two reasons: first, if the regime being monitored is not known, then one will not know to what sources to attribute high concentration measurements (*i.e.,* close sources or distant sources); second, the areal extent for which the observed concentration is "representative" will be impossible to determine (*i.e.,* measurements within the microscale regime may be representative of the concentrations occurring only within a few meters of the sampling station, while measurements within the mesoscale regime should be representative of the concentration occurring within hundreds of meters radius of the sampling site).

OPTIMIZING SPATIAL AND TEMPORAL RESOLUTION

The cost of purchasing and operating an air monitoring network will depend on the level of temporal and spatial resolution of air quality measurements desired. The requirements for good spatial resolution differ from the requirements for good temporal resolution. Good spatial resolution can only be obtained by sampling at numerous locations, while good temporal resolution can only be obtained by sampling over short intervals for a long duration. Achieving both spatial and temporal resolution requires monitoring at many sites, for a long duration. This becomes so expensive that most projects either "trade off" temporal resolution for spatial, or spatial resolution for temporal. Usually both types of studies must be conducted simultaneously; thus characteristics of both types must be incorporated into the air sampling program design. In most cases, the goal of the survey design is to achieve "acceptable" temporal and spatial resolution at the lowest cost. "Good" resolution is very expensive to obtain and is required only for research or where unique or complex air quality problems exist.

The concept of a study designed to achieve good temporal or good spatial resolution is best illustrated by the extreme case. For example, an air quality study design providing good spatial resolution might consist of 20 to 30 semi-automatic air samplers located at an equal number of different sites; all samplers would be operated simultaneously. Ideally, this program would be conducted so that measurements are taken under all wind directions and with all possible sources acting. The resulting data would show where the highest and lowest air pollution concentrations occur, and under what conditions (*i.e.*, wind direction, season of year, etc.). A study designed to provide good temporal resolution might consist of a single station using a continuous air sampling method. This station would be operated 24 hours a day, with averages each hour for an entire year. The results would provide the most probable and "worst case" diurnal profiles of pollution levels, the seasonal fluctuations and frequency of occurrence of worst case one-hour, eight-hour and daily average air pollution concentrations. This type of study would provide the maximum amount of information regarding the temporal distribution of air pollution concentrations, illustrating when, how frequently and how long adverse conditions occur at one location.

The approximate capital cost of the equipment required to conduct an air quality study is shown in Figure 36. The relative cost of using

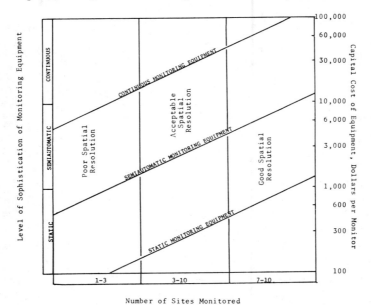

Figure 36. Achievable spatial resolution and capital costs for conducting an air quality survey with various numbers of monitoring sites and three levels of equipment sophistication.

static air samplers, is compared to the cost using semi-automatic or continuous air samplers. These curves allow one to determine either the approximate number of stations and the type of equipment which can be purchased, or the least cost number of sites and type of equipment that will fulfill the monitoring objective.

The three types of equipment indicated in Figure 36 represent three distinctly different levels of sophistication. The approximate 1975 costs of single static, semi-automatic or continuous air samplers are $50, $500 and $5000, respectively. These estimates represent the purchase price of the equipment only. The cost of shelters to house continuous monitoring equipment must be added. In most cases, these estimates are conservatively high.

The costs of operating stations for different durations is illustrated in Figure 37. These estimates assume that the operating cost for a continuous station is $25,000 per year. The cost of operating a semi-automatic station is approximately $2500 per year and a static station is $250 per year. In order to consider the additional expenses encountered with short-term projects (i.e., a few days) the cost per day is assumed to be four times the cost per day for a year-round station.

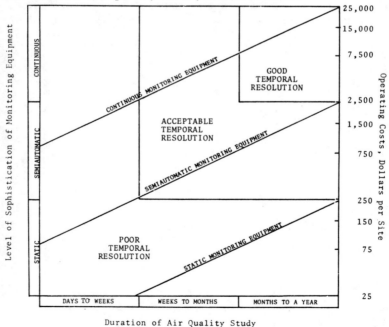

Duration of Air Quality Study

Figure 37. Achievable temporal resolution and operating costs for an air quality survey of various durations for three levels of equipment sophistication.

The estimates are for total costs to operate and reduce the data from a single station. Where appropriate, the cost estimates include the costs of operating and maintaining the monitoring equipment (including utilities, spare parts, chemical reagents, technician costs, etc.), calibration of the monitoring equipment, operation and maintenance of data recorders, and data reduction and logging. One of the assumptions upon which the cost estimates are based is that a continuous sampling station requires a considerable amount of "technician's time" each day, while a semi-automatic station requires no more than a single short visit each day, and a static air sampler will be visited only once per month. However, semi-automatic equipment operating on short sampling intervals may require considerably more of the technician's time. In such a case, the cost curve may approach that of a continuous station because a major portion of the cost of operating an air monitoring station is manpower.

CHAPTER V

SITE SELECTION FOR AIR MONITORING NEAR LINE SOURCES

SELECTING MICROSCALE AIR MONITORING SITES

In this chapter, definitive methods for selecting sites are presented, based on the air pollution dispersion models outlined in the California Air Quality Manuals[1] and the quantitative definition of microscale and mesoscale regimes given above. Whenever the physical environment to be monitored violates either of these assumptions, the site selection methods presented may not identify the optimum site.

Figure 38 is a graphical description of the microscale regime near a highway; the regime extends from the upwind edge of the highway downwind where the ambient pollution concentration approaches the local background concentration. The figure shows the mechanical mixing cell, as defined in the California Model,[1] where pollutants emitted by automobiles are thoroughly mixed resulting in a region where air pollution concentrations are uniform and relatively high. As the pollutants are transported and dispersed by the wind, the concentrations decrease as the distance from the highway increases. When the concentration approaches the local background concentration the boundary of the microscale regime has been reached.

The microscale regime can also be illustrated graphically by plotting isopleth lines of concentration levels as a function of the concentration in the mixing cell. Figure 39 is an example that shows lines of similar concentration downwind from the edge of a highway source in the horizontal direction, plotted on the abscissa, and the vertical direction, plotted on the ordinate. The family of curves plotted in Figure 39 illustrates the vertical and horizontal locations where the pollution concentration is reduced to 80%, 60%, 40%, 20% and 10% of the roadside edge levels.

75

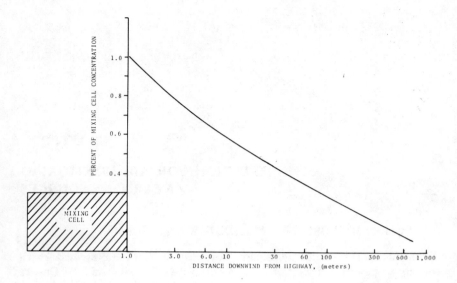

Figure 38. Decrease in ground level ambient air pollution concentration at increasing distance downwind from highway edge (according to California's model[1]). Assumes C stability and zero background concentration.

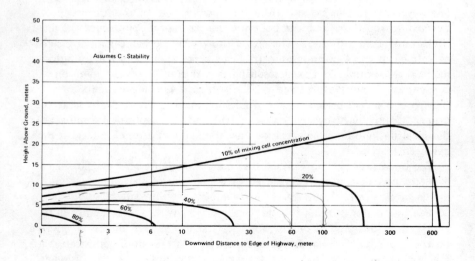

Figure 39. Isopleth concentration lines downwind of a highway line source (California's highway model, at-grade section[1]).

Figure 39 may be used to determine the optimum spacing for micro-scale monitoring stations used to collect data for validation of the California Life Source Model for the crosswind, C stability, ground level concentration case. Sites should be spaced so that the pollution gradient can be sampled at regular intervals. For a five-downwind-sampling site network, sites located at the mixing cell, and at the 80%, 60%, 40% and 20% isopleths would meet this requirement. The locations would be 0, 2, 7, 23 and 200 m from the downwind edge of the pavement.

All the curves plotted in Figure 39 were determined using the California Highway Line Source Model for crosswind conditions and C stability. This should provide optimum site spacing for ground level concentrations under C stability and crosswind conditions, but less than optimum for other conditions. In order to design the best possible site-spacing pattern, an evaluation should first be performed to determine the full spectrum of events likely to occur downwind of the source. This can be accomplished using historical meteorological data and diffusion models. Figure 40 illustrates the results of such an evaluation, showing the normalized ground level concentration gradient versus normal distance from the road for all possible conditions. Also given (in parentheses) is the frequency of occurrence of each condition of wind direction and stability taken from historical records for the time period (*i.e.*, season and time of day) during which sampling is proposed.

Since the precise conditions of meteorology that will be encountered in the field cannot be predicted beforehand, the sampling sites should be selected based on optimum spacing for the most probable stability and wind direction condition determined for historical records. Choosing sites using the most probable stability and wind direction should provide the greatest amount of data collected from optimally located sites. If optimum spacing of sites is desired for measurements under a different condition of meteorology, particularly the worst case condition of dispersion, additional sampling sites may be added in order to supplement the coverage provided by sites already chosen. In this way the sampling array is designed to provide the best data under specific meteorological conditions (*e.g.*, most probable and worst case) while providing less than optimum coverage for other conditions likely to be encountered during the study.

The site selection procedure is best described by use of an example such as Figure 40. The sites shown in Figure 40, located at distances 1, 2.3, 7.2, 33 and 260 m from the road, were chosen in order to allow sampling during crosswind conditions at regular intervals of 100%, 80%, 60%, 40% and 20% of the roadside "mixing cell" concentration, C_{mc}, within a tolerance range of ± 10% of C_{mc}. Using a 10% tolerance range,

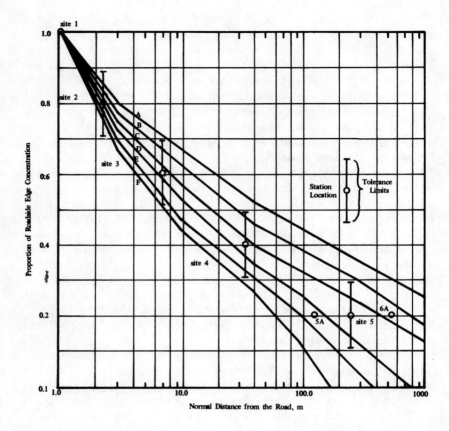

Figure 40. Optimum locations of air sampling sites determined for crosswind conditions according to the California model.[1]

a site objective of 20% C_{mc}, for example, is considered met whenever the concentration is between 10%-30% C_{mc}. This tolerance range is shown graphically in Figure 40 by a vertical line centered at each site and crossing concentration lines predicted by the model. If the tolerance range crosses a concentration line, then the site can be expected to meet its intended sampling objective within ± 10%. Figure 41 shows the concentration profile lines predicted for both crosswind and parallel wind conditions and hypothetical frequencies of occurrence of each condition, which can be determined from historical meteorological records. By summing the frequency of occurrence of each concentration profile line

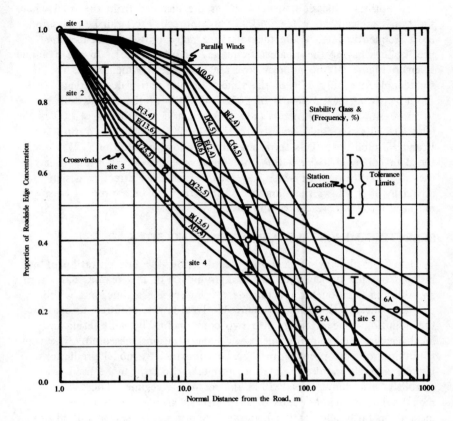

Figure 41. Relative concentration versus normal distance from an at-grade highway for all wind directions and stabilities calculated from the California model.[1]

crossed, the frequency at which the monitoring site will meet its intended sampling objective can be determined. For example, the site locations shown in Figure 41 should meet their intended objectives 100%, 85%, 81.6%, 78.2% and 51% of the time for sites 1-5 respectively. However, these sites do not provide uniform coverage at regular intervals for parallel wind conditions, measuring at distances where the ambient concentration is roughly 100%, 97%, 89%, 53% and <1% of the mixing cell for sites 1-5, respectively. If model validation under parallel winds is desired then additional sites may be needed to measure at distances away from the road where concentrations are reduced to $\cong 70\%$ and $\cong 25\%$ of C_{mc}.

The optimum location of each site is the distance from the road where the target concentration (*i.e.,* 40% of mixing cell, etc.) can be measured most frequently (*i.e.,* where the tolerance range crosses concentration profile lines having the greatest frequency of occurrence). In the example given in Figure 41 the poorest coverage is at site 5, which meets its stated objective only 51% of the time. An alternative design that should provide measurements of 20% C_{mc} 78% of the time would be to move site 5 to 150 meters (site 5A) and add a sixth site at 600 m (site 6). Site 5A could be used to measure $20 \pm 10\%$ C_{mc} during D, E and F stabilities. This sixth site should allow an additional 27% of the data collected to be within the stated objectives at all sites. The use of a sixth monitoring site may be the most cost effective design, especially if it allows the duration of the study to be shortened because of the improved coverage of the sites.

SELECTING MESOSCALE AIR MONITORING SITES

Data from downwind sampling sites describe the concentration of pollution from the project plus the background. For this reason, data from an additional station located outside the microscale regime, but within the same mesoscale regime is required. This data is required to separate the pollution concentrations into two components, by subtracting the local background from the microscale concentrations measured. The contribution from the project can then be compared to model predictions. Usually a single monitoring station, properly chosen, can be used to establish background levels within the mesoscale regime. Microscale model validation experiments often utilize an array of air samplers located on both sides of the highway. As long as winds do not blow parallel to the roadway, samplers on one side of the highway will provide background data, while samplers on the other side of the highway will measure the highway contribution.

Figures 39 and 40 can also be used to determine a good location for a mesoscale site. The isopleth lines can be used to estimate the theoretical boundary separating the microscale regime of a nearby highway from the mesoscale regime. Once the boundary has been determined, stations designed to sample background concentrations can be located outside the boundary, while stations designed to sample pollutants from the highway plus background can be located within the boundary.

The isopleth that represents the boundary separating the microscale and mesoscale regimes depends on both the concentration within the mixing cell and the background concentration. Figure 42 provides a graphical method for determining which isopleth to use when sampling

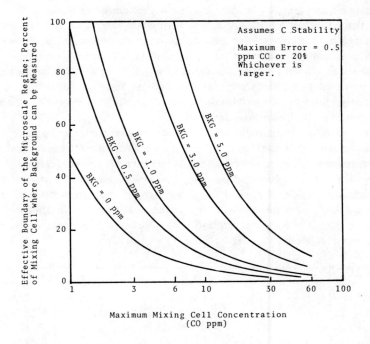

Figure 42. Percentage of mixing cell concentration where background concentrations can effectively be measured (function of mixing cell concentration and probable background level).

for carbon monoxide. The equations used in Figure 42 are:

$$\% \ C_{mc} = \frac{0.5 - BKG}{C_{mc}} \times 100 \ \text{for} \ C_{mc} \leqslant 2.5 \ \text{ppm CO} \tag{1}$$

or

$$\% \ C_{mc} = \frac{0.2 \ C_{mc} - BKG}{C_{mc}} \times 100 \ \text{for} \ C_{mc} \geqslant 2.5 \ \text{ppm CO} \tag{2}$$

This figure can be used to determine what percentage of the mixing cell concentration, $\%C_{mc}$, when added to the background concentration, BKG, will increase the background concentration by 0.5 ppm of CO or 20% (whichever is larger). This means that an air monitoring site located outside the designated isopleth (measured in percentage of mixing cell concentration) will measure the local background concentration with an error of up to 20% or 0.5 ppm CO (whichever is larger).

Using Figures 39-42 one can determine the minimum distance down-wind from the highway where a monitoring site should be located for measuring background levels. Mesoscale sites can be located at greater distances downwind or upwind as long as they are outside the microscale regime of the highway. Stations located upwind should also be located at greater distances than the minimum, because the wind direction changes frequently throughout the day and a sampling site that is upwind during one hour may be downwind the next.

There is also a maximum distance within which the sampling site should be located if there is another major source of pollutants nearby. This maximum distance depends on the distance to the source (see Figure 43). If a background sampling site is located too far away from the highway corridor for which local background measurements are desired, the measurements taken at the distant site may be significantly different from the local background concentrations that occur at the corridor under study.

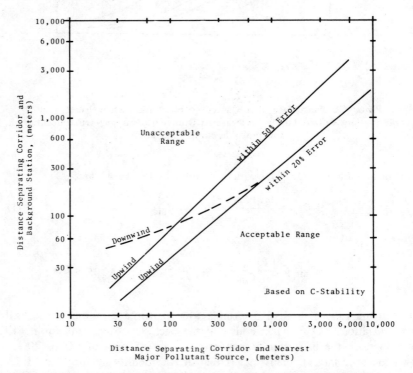

Figure 43. Maximum distance from project corridor that a mesoscale background station should be located versus distance to nearest major pollution source.

Figure 43 illustrates the maximum distance that a background site should be located from the highway corridor when other large sources are nearby. The limiting value is based on a maximum error of 20% between the value measured at the distant air sampling site and the expected local background concentration at the highway corridor.

Figure 43 is to be used in selecting a location for a mesoscale site when there is a heavily traveled freeway or arterial within 1000 m of the corridor under study. Under these conditions the maximum distance from the corridor and the minimum distance determined using Figures 39-42 may approach similar values. This indicates that the location of the sampling station should be selected precisely, in order to obtain accurate results.

Besides the ground level sampling site locations discussed above, local background air pollution concentrations can be measured from air sampling stations located on the top of buildings. Mesoscale sites located on the top of buildings can provide useful and accurate information. However, the buildings must not be too tall or the measurements will not represent the background concentration at ground level. Pollution from nearby sources may contribute significantly to the ground level background concentrations, but not to those measured atop a tall building. For this reason, the height of the sampling site must be limited.

As in the case of the maximum horizontal distance away from a project corridor, the maximum height for a mesoscale monitoring site depends on the distance from the closest major source. Figure 44 illustrates the

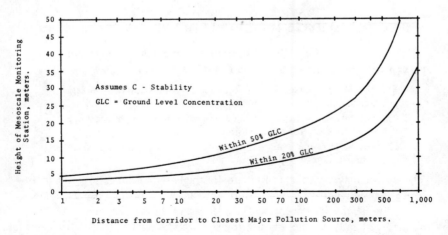

Figure 44. Acceptable height of a mesoscale monitoring station if ground level concentrations are to be measured.

relationship between distance to a contributing source and the percentage of ground level air pollution concentrations occurring at different heights. With this figure the maximum height at which the local background concentrations can be measured (within an accuracy of 20% of ground level background) can be determined. This error estimate is conservative as it assumes that the closest major source contributes 100% of the ground level concentration and 80% at the indicated height. In practice, no single source would be responsible for all of the local background pollution concentrations, because sources located at greater distances from the sampling site contribute more uniformly to both the concentration at ground level and atop a building. Therefore, in practice, the error between the elevated and ground level measurements will be smaller than 20%.

Figures 39, 43 and 44 were developed using the California Highway Line Source Model for at-grade highway sections, crosswind conditions, and atmospheric stability class C. Stability class C was selected as the stability class most frequently encountered under daytime conditions with urban terrain. Clarke and McElroy[2] have reported measurements of a relatively unstable boundary layer over urban terrain even when adjacent rural environs exhibit very stable (stagnation) conditions near ground level. Clarke and McElroy's results tend to reinforce the use of C stability as the most probable condition encountered within urban environs. If different stability classes are desired, or different models are to be tested, then similar figures can be developed for selecting optimum air monitoring sites.

WHERE TO VALIDATE MICROSCALE MODELS

To obtain data for validating microscale models for environmental assessment, monitoring sites should be chosen where:

1. The highest concentrations from the project occur (due to large traffic volumes or narrow right-of-way),
2. The highway configuration and upwind topography is most representative of the whole project, or
3. The basic assumptions of the model are violated (due to highway configuration or topography).

The model may be validated where the highest concentrations occur because that is where the most confidence in the model is needed. Model validation at the most representative location allows the model to be applied generally to the whole project. Validating the model for irregular terrain (hills, valleys or nearby tall buildings) or complex highway configurations (intersections, elevated or depressed sections, and unusual lane configurations) may be required since models are the least reliable when basic assumptions (*i.e.,* smooth level terrain, uniform wind flow field and wind speeds greater than 1 m/s) are violated.

Once the sites have been selected, the highway route should be monitored in cross-section sampling at various horizontal or vertical distances from the highway. Special attention should be given to measuring the air pollution concentration at the right-of-way edge. Background concentration measurements should be subtracted from downwind measurements to determine the contribution due to the highway.

DURATION OF A MICROSCALE STUDY

The sampling survey should be conducted for a sufficiently long duration so that observations are made under a wide range of source-related and meteorology-related conditions. Typically, if meteorological conditions are favorable (*i.e.,* a wide range of wind directions, wind speeds and atmospheric stabilities), an intensive air monitoring study lasting from 1-3 weeks is sufficient for microscale model validation. In most cases, air samples should be collected using 15-60 minute averaging times, but sampling for shorter averaging times can be used if traffic and wind data are gathered for comparable intervals. Averaging samples for time periods longer than one hour produces additional errors because changing wind direction has a nonlinear effect on the downwind concentration. Whenever the wind direction is not persistent, the short averaging time samples (*i.e.,* 15-minute averages) may be the best.

Air sampling may be conducted either during peak hour traffic or 24 hours a day. For model validation, it is desirable to sample during each hour of the day and night that there is enough traffic to produce measurable pollutant concentrations. In this way, a large amount of data can be collected over a short period of time to allow validation of the model under different meteorological conditions.

The duration of study required can be estimated using historical meteorological data and the statistical methods given in Chapter XV. By way of model calculations, the average and range of concentrations expected to be measured at each station can be determined. The range divided by about six can be used to estimate the standard deviation. Dividing the range by six assumes that the extreme values calculated approximate the highest and lowest 0.1% of all possible concentrations. Using the nomographs in Chapter XV, the number of samples needed can be estimated. The duration of study required can then be determined by dividing the number of samples required, n, by the frequency of occurrence of a particular condition of meteorology expressed as a percentage, f, times the number of samples, S, collected at each site per day or hour.

$$\text{length of study required} = \frac{n}{f(S)} \tag{3}$$

Typically, values of n range from 30-100 samples, and S might be from 1-4 per hour. For unusual events occurring less than 5% of the time and if 40 samples were needed and S = 4 per hour, then the duration of study required would be 200 hr. At 12 hours a day of sampling, this would require more than two weeks of sampling. For frequently occurring conditions, sufficient data might be obtained in a few days.

REFERENCES

1. Beaton, J. L., A. J. Ranzieri, E. C. Shirley and J. B. Siog. *Air Quality Manual, Vol. IV, Mathematical Approach to Estimating Highway Impact on Air Quality* (Sacramento, California: Department of Public Works, Division of Highways, April 1972).

2. Clarke, J. F. and J. L. McElroy. "Effects of Ambient Meteorology and Urban Morphological Features on the Vertical Temperature Structure Over Cities," presented at the 67th Annual Meeting of the Air Pollution Control Association, Denver, Colorado, June 9-13, 1974, Paper No. 74-73.

CHAPTER VI

SITE SELECTION FOR
AIR MONITORING BACKGROUND CONCENTRATIONS

INTRODUCTION

The methods presented in this chapter for selecting monitoring sites for area sources are designed primarily for use in determining background concentrations near major point and line source projects. Site selection is then constrained to determining the maximum concentrations due to area sources within the major impact area of the point or line source under study. Worst case air quality levels most likely occur where high background concentrations are added to high project concentrations.

WHERE TO MONITOR

The number and location of sites where background air pollution levels should be monitored will depend on the variety of homogeneous areas within the overall study area. Homogeneous areas can be selected on the basis of land use patterns, common meteorological regimes, types of highway design and sensitive receptor areas. Types of homogeneous areas that may require monitoring are:

1. Land Use Patterns
   a. Forested areas
   b. Agricultural areas
   c. Urban areas
   d. Industrial areas
2. Meteorological Regimes
   a. Effects of topography
   b. Localized drainage winds
   c. Prevailing wind directions
   d. Wind speeds that produce maximum concentrations

3. Sensitive Receptor Areas
   a. Schools
   b. Hospitals
   c. Parks for children's activities
   d. Convalescent homes
   e. Residential areas

These features should be considered and may be plotted on a map of the area. Ideally, samples should be taken within each of the different homogeneous sections. However, if manpower and/or equipment limitations restrict the number of sampling sites, the sites to be air-monitored should be selected on a priority basis. Three locations have particular priority for environmental assessment (see Figure 1, chapter I): (a) where the maximum project concentrations, C_{max}, are expected; (b) where maximum background, $C_{B\,max}$, is expected; and (c) where high air pollution concentrations are expected at sensitive receptors.

Determining where maximum project concentrations, C_{max}, occur can be accomplished by use of an appropriate line or point source dispersion model. In the case of surveillance of a single major source (point or line) the location of $C_{P\,max}$ and the spatial distribution of air pollutants is primarily dependent on meteorology. For area sources, with ground level emissions, the spatial distribution of conservative air pollutants (*i.e.*, not photochemical) is primarily dominated by the distribution of sources because the effects of meteorology are the same for all sources. The one exception is the accumulation of pollutants in the downwind direction, where C_{max} occurs downwind of the greatest continuous area source extent.

Where maximum background concentrations are expected can also be determined using a mesoscale air pollution dispersion model. Otherwise, the best procedure may be to use a grid method of identifying where pollution emissions are high, and subsequent concentrations will probably be high. The procedure for using the grid method is to:

1. Locate the project area on a topographic map.
2. Identify probable homogeneous meteorological regimes based on uniformity of topography.
3. Divide each meteorological regime into areas of homogeneous land use.
4. Divide each land use area into uniform grid squares (see example for a proposed highway corridor, Figure 45).
5. Calculate the hourly or daily average ground level air pollution emission flux (in μg-m^{-2}-sec^{-1}) for the grid squares from available emission inventories or traffic data.
6. Use the emission flux values as an indication of where monitoring stations may be appropriate.

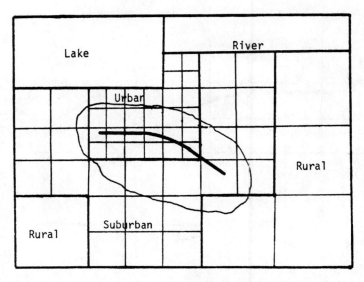

Figure 45. Study area divided into grid squares according to land use.

Homogeneous meteorological regimes are assumed to prevail except where ground surface features affect the wind flow field. Surface features such as mountains, ridges, valleys, coastlines or large lakes, and divergent urban and rural land forms should be noted as possible boundaries separating homogeneous meteorological regimes. Within each meteorological regime, areas of homogeneous land use can be divided into uniform grid squares (see Figure 45). The size of the grid square can vary; however, grid squares smaller than one square kilometer are not recommended. As shown in the figure, areas with low, uniform emission rates can be represented with larger grid squares. Areas where emission rates are relatively high and variable, as in downtown areas, may be better represented by small ones.

After the grid squares have been labeled according to emission flux (see Figure 46), the general locations where high and low concentrations occur can be inferred. Gifford and Hanna[1] have developed a simplified urban air pollution model employing uniform grid squares. They consider the accumulation of pollutants as the wind passes over a city to be a "weak function of city size." For quite large grid areas, 5 x 5 km or more, where the emission flux varies little from grid to grid, the grid square containing the receptor point is "as a rule the dominant one." For this reason, Gifford and Hanna conclude that it may be unnecessary to include the influence of remote upwind source areas.

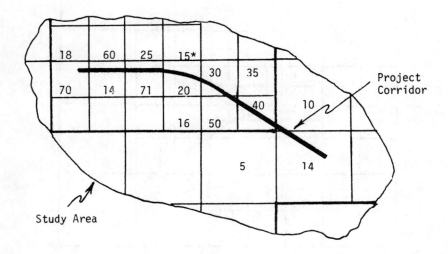

Figure 46. Average emission flux values for each grid square near project corridor. Numbers represent average area source emission densities of CO in $\mu g/m^2$-sec.

Grid squares labeled according to emission flux can be used to determine where the highest pollution concentrations are likely to occur. Using grid blocks such as shown in Figure 46, it is possible to estimate the location of the maximum background concentration, $C_{B\,max}$. A monitoring station near the center of the maximum grid square should provide reasonable results. The highest concentrations will occur within the highest grid, probably toward the downwind side. An additional station near a downwind edge of the grid may be desirable. When good spatial resolution of background is needed, stations should also be located within grid squares having moderate and low emission fluxes.

Locating monitoring stations near sensitive receptors is as important as measuring the maximum background concentration. Maximum background is needed for "full disclosure" in environmental impact statement; however, the impact at sensitive receptors must also be documented. Some data is needed to assure that levels at sensitive receptors will be lower than levels at maximum background sites, especially if levels are expected to be near the NAAQS (National Ambient Air Quality Standards), or if controversy is anticipated. Whenever possible, some amount of monitoring should be conducted at each sensitive receptor site along the project corridor. Sensitive receptors may, of course, be selected for physiological or political reasons.

Once the general locations for mesoscale monitoring sites have been selected, the specific location of each site should be determined using the information presented in Figures 39-44 to insure that the site is within the appropriate mesoscale regime and outside any microscale regimes. Care should be taken to insure that each sampling site meets the additional criteria of

    (a) area wide "representative exposure,"
    (b) accessibility for personnel,
    (c) availability of electrical facilities (if needed), and
    (d) provisions of some degree of equipment security (storm damage and vandalism).

NUMBER OF MESOSCALE SITES REQUIRED

A survey designed to provide good spatial resolution for background levels within the mesoscale regime must have the "right" number of stations, at the "right" location. The number of sites needed will depend on the diversity of land use, the meteorology, the number of sensitive receptors, and the degree of spatial resolution desired. Figure 36, page 71, illustrates the relationship between the number of stations, the level of equipment and sophistication employed at each station, and the expected quality of the resulting study in terms of the ability to provide spatial resolution. The three zones designated as "poor," "acceptable" and "good" estimate the spatial resolution expected for the experiment. Use of only a few stations will provide "poor" spatial resolution, because the locations where minimum, maximum and average concentrations occur cannot be clearly separated and identified. A study achieving "good" spatial resolution requires simultaneous measurements from many locations to establish the area wide mean concentration while identifying locations where the minimum and maximum concentrations occur. "Acceptable" spatial resolution requires several carefully selected stations, which provide sufficient evidence that the location where the highest concentrations occur has been identified and monitored, and that the levels everywhere else within the study area exhibit lower pollution concentrations. Under conditions of moderately diverse land use, a survey designed to provide an "acceptable" level of spatial resolution may require from 5 to 15 sites. If land use patterns are very diverse, more sites may be needed.

AVERAGING TIMES FOR THE AIR SAMPLES
AND DURATION

An air quality study should provide adequate temporal, as well as spatial, resolution of air pollution concentrations. An air sampling survey designed to provide good temporal resolution within the mesoscale regime must: (a) include samples having a short enough averaging time to compare to one-hour air quality standards; and (b) last long enough for the effects resulting from all possible combinations of source strength and meteorological phenomena to be observed or at least reliably estimated using statistical techniques. Obtaining samples averaged over a short interval requires either continuous sampling, or grab samples collected over one-hour intervals. This eliminates several methods of air sampling which collect samples averaged over much longer time periods. Monitoring long enough to sample all variations of source strength and meteorology may require as long as one year.

Continuous monitoring for one year assures that samples are taken during all seasons of the year, on all days and during all hours. Such sampling would provide the best data describing the temporal distribution of pollution concentrations. However, the high cost of maintaining and operating such a station, as well as the long lag time between the initiation of the survey and the final results, are unreasonable. Therefore, most surveys must compromise the best available methodology by using a sampling schedule designed to include only the periods when the highest pollution concentrations are expected to occur. In this way the "worst case" pollution levels can be determined while the temporal distribution of low concentrations will be less accurate. Such a sampling program would tend to report "conservatively higher" results. This compromised survey would be a "worst case" analysis, and usually would represent an acceptable approach for assessing environmental impact.

In order to conduct a "worst case" analysis, periods of high pollution concentration must be predictable. When designing sampling schedules, one must know when the highest concentrations are likely to occur and schedule accordingly. In most urban areas, these time periods are regular and reproducible. Some air pollutants exhibit "worst case seasons" during the year, "worst case days" during the week and "worst case hours" during the day. For example, the worst case season for ozone might be summer; the worst case days might be weekdays; and the worst case hours might be from 12 noon to 7 p.m. A survey designed to provide a "worst case" analysis for ozone at such a station would only require monitoring between 12 noon to 7 p.m. each weekday during the summer months. Compared to a full year study, this is a considerable reduction in the amount of monitoring required. Figure 47 illustrates hypothetical temporal profiles of ambient CO and O_3 air pollutant concentrations with typical seasonal, daily and diurnal air pollution trends. Specific profiles can be developed with historical data available from CAMP stations or state and local air pollution control agencies.

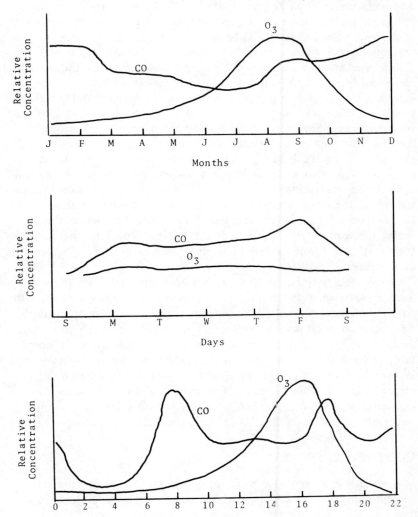

Figure 47. Hypothetical profiles of typical seasonal, daily and hourly fluctuations in CO and O_3 concentrations.

Seasonal, daily and diurnal pollution profiles vary for different pollutants, different geographical areas (macroscale variations) and at different locations within a study area (mesoscale variations). This is due to variations in source strength and meteorology. The most reliable method for predicting the worst case periods is to use historical air pollution measurements from an air monitoring station operated by the local Air Pollution Control Department (APCD) and located within or near the study area. The highest pollution concentrations should correspond to the periods of peak concentrations measured at the APCD stations. This is true only when the APCD station is located within the same mesoscale regime as the study area. If the APCD station measures pollutants within a mesoscale regime or is located very distant from the study area, there can be large differences in the occurrence of maximum concentrations. Similar temporal profiles in ambient pollution level cannot necessarily be inferred. For example, if the APCD station is located near the south edge of an urban area and the study site is near the north edge, maximum background concentrations at the APCD will occur when winds blow from the north; conversely, maximum concentrations at the study site will occur when winds blow from the south. In this example, the worst case season for APCD station might be during the winter when the prevailing winds are out of the north, while the worst case season for the study site might occur during the season of the year when the prevailing winds are southerly.

Peak concentrations can also be "out of phase" with APCD peak measurements when the study site is located several kilometers from the APCD station. For example, in an urban area photochemical oxidant concentrations may peak at midafternoon, and this same air mass might then be carried over rural terrain. The ground level oxidant concentration in the rural area would peak several hours later in the afternoon.

These examples illustrate that historical data can be misused for predicting worst case time periods. Historical data should be used, but judiciously, taking into account the effects of wind speed, wind direction and relative location of the APCD station and the study site.

SAMPLING SCHEDULES

Once the most probable worst case periods have been estimated, a sampling schedule can be prepared that includes all these worst case time periods. The duration of the sampling survey should be increased by 35-50% to assure that worst case conditions are observed. This additional sampling should be conducted half before and half after the expected worst case time period. For example, when the worst case season is estimated between mid-December and the end of January, an additional ten days of sampling during the first weeks of December and the first weeks of February would provide a safety factor to assure that worst case concentrations are included. Typically, background carbon monoxide studies should be conducted for one to six months including the worst case season of the year.

When employing a sampling network consisting of primary and secondary sampling stations, sampling schedules for secondary stations may require sampling only on some of the days that the primary station is operated. Choosing which days can be done systematically (*i.e.,* every other day, or every third day), or randomly (*i.e.,* using a random number table). An unbiased sampling schedule should be prepared at the beginning of the sampling program and followed to its conclusion. The primary stations should be sampled at least every third day to assure that no major meteorological phenomenon, such as high-pressure related subsidence inversions or stagnations, are entirely missed.

The number of days that sampling is required can be estimated using Figures 143 and 144 in Chapter XV. For example, if the predicted standard geometric deviation of daily average SO_2 concentrations is between 2.0 and 3.0 and the worst case season can be predicted within 90 days, the number of days that require sampling in order to determine the daily mean SO_2 concentration within ± 20% with 95% confidence is from 34 to 56 days (derived from Figure 144, page 278). The sampling days should be evenly distributed over the 90-day period. The specific days can be chosen on either a systematic or a random basis.

REFERENCES

1. Gifford, F. A. and S. R. Hanna. "Modeling Urban Air Pollution," *Atmos. Environ.* 7:131-136 (1973).

CHAPTER VII

SITE SELECTION FOR AIR MONITORING NEAR POINT SOURCES

INTRODUCTION

This chapter describes a general methodology for selecting air monitoring sites near point sources. The objective of monitoring a point source is assumed to be to determine the maximum impact of the source on air quality in the immediate area. Most important are the maximum ground level concentrations, which occur for various averaging times (*i.e.,* 1-hr, 3-hr, 1-day and 1-yr). The basic approach taken here is to determine where the maximum concentrations are expected to occur using point source atmospheric diffusion models, then locate monitoring sites in these areas. Measured concentrations can then be compared to NAAQS.

The methodology recommended includes six major steps: (a) an evaluation of historical meteorological records, (b) setting precisely defined monitoring objectives, (c) running models to determine C_{max} and where it most likely occurs, (d) determining the size of the representative sampling area that can be covered by a single station (coverage ratio), (e) determining the minimum number of sites needed to insure a desired number of observations of C_{max} (a statistical approach), and (f) the optimum number of location of stations required to meet monitoring objectives with a high probability of success.

EVALUATION OF HISTORICAL DATA

It is very important before choosing a monitoring site to evaluate the type and frequency of meteorological conditions likely to occur, especially conditions adverse to the dispersion of air pollutants. This includes wind speeds, wind directions, atmospheric stability conditions, mixing heights, vertical temperature gradients and any other available data pertinent to atmospheric diffusion. Ground level pollution concentrations near point sources are highly dependent on meteorological conditions, which affect plume rise and the rates of transport and diffusion of pollutants.

There are a number of meteorological phenomena that produce especially high ground level pollutant concentrations. Some of the most important conditions are inversion breakup fumigation, plume trapping by an elevated inversion and coning under critical wind speeds. These conditions usually produce the highest concentrations but occur infrequently. Other equally important conditions are those that produce near maximum concentrations and occur much more frequently (such as stable and unstable conditions with low wind speeds). It is also valuable to determine the meteorological conditions that occur most frequently (*i.e.,* annual mean wind speed and neutral conditions) since these conditions may affect long term average concentrations. Before selecting air monitoring sites the meteorological conditions for which monitoring data is desired must be specifically delineated. Locations where the maximum concentrations occur under these conditions can then be determined using diffusion models.

USING MODEL RESULTS TO DELINEATE POTENTIAL ZONES FOR MONITORING

When selecting air monitoring sites near a point source it is highly desirable to have a good idea where these maximum concentrations are likely to occur so that the monitoring station can be so located. At the present time the best method of determining where the maximum concentrations occur is to use point source diffusion models. The diffusion of air pollutants from point sources is quite complex due to the number of variables interacting to affect nearby ground level concentrations (*i.e.,* wind speed, wind direction, atmospheric stability, inversion heights, plus emission rates, plume rise and topographic effects). Therefore, the meteorological conditions that produce maximum pollution levels are not always easy to determine. However, an organized step-by-step approach can be used to determine the locations where maximum levels occur and the likelihood that an air monitoring station located at these points would in fact measure the maximum values.

Because sampling objectives may include monitoring under many meteorological conditions, a large number of sites may be required. A good method of selecting potential zones for monitoring is to display the results of diffusion calculations as shown in Figure 48. This figure shows graphically the range of distances downwind from the source where

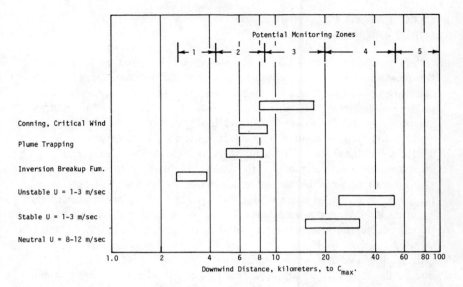

Figure 48. Identifying potential zones where C_{max} may occur.

maximum concentration occurs under each range of meteorological conditions. This figure can then be used to select potential sampling sites.
For example, a site located in zone 1 would be needed to monitor
during low wind speed and unstable conditions; zone 2 is required for
monitoring during both trapping and inversion breakup conditions;
zone 3 could monitor during coning at critical wind speeds and mean
wind speeds under neutral conditions; zone 4 could monitor both mean
wind speed-neutral conditions and low wind speed-stable conditions; in
some cases sites in zone 5 may be desirable to document that ambient
levels are less outside the major impact area described by the inner zones.

Figure 48 becomes even more valuable when calculated values of
maximum concentration, C_{mc}, are recorded in each box as well as the
probable frequency of occurrence of various wind speeds, stabilities,
mixing heights and wind directions that can be obtained in summary
form from the National Weather Service.[1]   These data can then be
used to establish priorities in choosing monitoring sites.   Monitoring
locations where maximum concentrations occur very frequently should
probably receive a higher priority than locations where C_{max} occurs
infrequently.

CALCULATING C_{max} AND X_{max}
USING MODELS

When using diffusion models to assist in air monitoring site selection, the most important result obtained is where the maximum ground level concentrations, C_{max}, occur. For a point source C_{max} occurs along the plume centerline at a distance, X_{max}, from the source. Several methods are presented to determine C_{max} and its location downwind, X_{max}.

The first step in point source modeling is to calculate the plume rise. This can be done using TVA's nomograph presented in Chapter II, Figure 26, page 40. Using Figure 26, the plume rise under either neutral or stable conditions can be estimated for various wind speeds. This value for plume rise, Δh, plus stack height, h_s, equals the effective stack height, h_e, used in diffusion calculations. Other methods of calculating the expected plume rise have been presented by Briggs[2] and Fay et al.[3] and can be used in place of Figure 26 if preferred. Most authors agree that estimating plume rise under unstable conditions is questionable at best due to the lack of data and uncertainties in defining air motions under unstable conditions. For this reason a conservative solution is recommended that is, to underestimate plume rise by using TVA's[4] nomogram for neutral conditions and wind speeds equal to 3 m/sec. (3 m/sec is TVA's reasonable minimum $\bar{u}$ at stack top.)

The next step is to determine the distance downwind where the maximum concentrations occur under each meteorological condition of interest. This can be determined for the coning plume, trapped plume, and inversion breakup fumigation by using nomograms prepared by TVA shown in Figures 30 and 31, pages 45 and 46. Another method, which is especially appropriate for short to moderately tall stacks where inversion breakup fumigation and plume trapping are not problems, employs a figure prepared by Turner[5] (see Figure 29, page 43). By first using Figure 26 to determine the plume rise, Figure 30 can then be used to determine the distance downwind where the maximum concentration occurs. This figure is based on the Pasquill-Gifford[6] curves for σ_y and σ_z and is also convenient for calculating C_{max}.

Each of the methods described for locating potential monitoring sites gives precise results implying that the best location is at exactly the distance calculated by the models. In fact, diffusion calculations as recommended above do not predict accurately the precise location where maximum concentrations occur. A more realistic solution to the models may be to calculate where maximum concentrations occur for a narrow range of conditions rather than a precise point. For example, rather than determine where the maximum concentration occurs for neutral stability at the mean wind speed, $U = 5.7$ m/sec, it may be more realistic to determine X_{max} for wind speeds for say 4-7 m/sec. X_{max} would occur at different locations for $U = 4$ m/sec due to the difference in plume rise. The frequency of occurrence of wind speeds between 4-7 m/sec could be determined from historical meteorological summaries such as STAR.[7]

DETERMINING THE OPTIMUM NUMBER AND LOCATION OF SITES

When air monitoring near point sources the number of sites needed to confidently determine the total impact of the source can be quite large. As shown in Figure 48, four or five zones may require data in a single direction downwind. If this number is multiplied by the number of directions from which the wind can blow, then a very large number of monitoring sites may be required. In practice, such a large number of sites is usually not economically acceptable. Therefore, the survey objectives are compromised by monitoring in only a few selected downwind directions. Which downwind directions should be monitored can be determined from an examination of historical wind direction/frequency data, and a consideration for the potential receptors in each direction.

Realistically, maximum ambient air pollution concentrations can never be measured precisely because of the extremely large number of sampling locations that would be required to define the entire concentration profile and identify the maximum. Even when diffusion models are used, they predict that maximum concentrations, C_{max}, occur at a single point on the ground; however, by using a tolerance range about C_{max} an area can be defined within which the ambient concentration is equal to C_{max} minus the tolerance range. Monitoring locations can then be chosen within this defined area for the purpose of measuring maximum concentrations within some tolerance (i.e., within 10% of C_{max} or within 20% of C_{max}, etc.). The acceptable tolerance range will have an effect on the number and location of stations for air monitoring; the greater the tolerance, the fewer stations that will be required. An acceptable tolerance range should be based on analysis of the maximum expected levels compared to air quality standards. If maximum levels are expected to be near standards then the tolerance should be small. Conversely, low concentrations need not be resolved as accurately as high concentrations, so that a greater tolerance range may be acceptable. The example calculations and figures presented in this paper assume a tolerance range of 10%. Similar figures can be prepared for other tolerance values.

Before selecting specific sites for air monitoring within each zone it is necessary to provide general solutions of Gaussian point source dispersion models. From these general solutions conclusions may be drawn regarding the proper location and spacing of sampling sites.

REPRESENTATIVE SAMPLING AREA

One of the most important determinations that can be made using the models is the size, shape and extent of the representative sampling area about a monitoring station. Representative areas can be defined using a tolerance range in the same manner as the areas about the point where C_{max} occurs were defined. Knowing the size and shape of the representative area allows guidelines for minimum spacing between stations and assists greatly in determining the number of stations required.

From Turner's[5] curves (see Figure 27) of normalized concentration, $X U/Q$, versus distance downwind, x, it is apparent that the point where the maximum ground level concentration occurs is also where the rate of change in the concentration with respect to downwind distance is at a minimum. This means that concentrations are fairly uniform where C_{max} occurs. Turner's curves can be used to determine the distance toward and away from the source within which the concentrations deviate from C_{max} only by some minimum amount (such as 10%). Figure 49, determined graphically from Turner's curves, shows the radial distance from the source, X_{10}, within which concentrations are within 10% of C_{max}, when the peak occurs at X, for two stability classes.

Besides being used to determine the extent of the "representative area" in the radial direction, Figure 49 also provides at least an upper limit to the number of monitoring sites needed in this direction. Even if sites are to be located to provide complete coverage so that at least one station monitors concentrations within 10% of C_{max}, the maximum number of stations required can be determined graphically as shown in Figure 50. Figure 50 illustrates the same boundaries shown in Figure 49 except that each "step" represents the area that can be covered by a single monitoring site. For example, if all C_{max} values are expected to occur within 1-10 km, then, from Figure 50, the maximum number of downwind stations needed in one direction is 4-5. Note that under unstable atmospheric conditions the boundaries in Figure 49 are narrower such that more "steps" are required to cover the same distance shown in Figure 50.

In addition to the downwind extent of the "representative area," it is also important to know the crosswind extent. The distance, Y_{10}, within

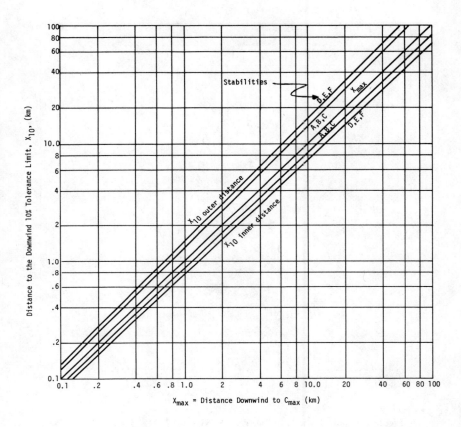

Figure 49. Graphical method for determining X_{10} versus distance downwind to X_{max} (X_{10} = distance from the source to 0.9 C_{max} = $|X-X_{max}|$).

which concentrations are within 10% of C_{max} can be calculated from the equation

$$(1-0.10) = \exp - \frac{1}{2} \left(\frac{Y_{10}^2}{\sigma_y^{\,2}}\right) \tag{1}$$

which is derived from the general Gaussian plume Equation 11, Chapter II. This reduces to

$$Y_{10} = 0.92 \; \sigma_y(x). \tag{2}$$

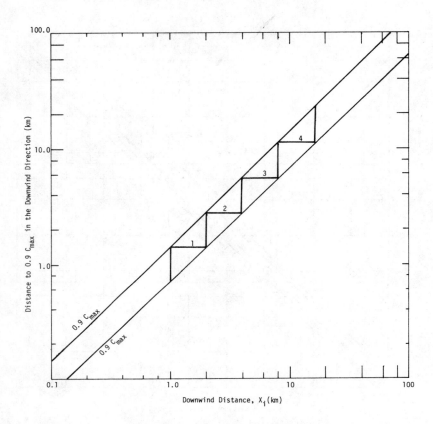

Figure 50. Illustrating the number of sites needed to maximize coverage between 1-10 km for D stability.

Values of Y_{10} for various distances downwind, and stability classes are plotted in Figure 51 (σ_y's are Pasquill-Gifford values).

Values of X_{10} and Y_{10} can now be calculated to determine the size and shape of the "representative area" under C_{max} conditions. As is apparent from a comparison of Figures 50 and 51, values of X_{10} tend to exceed Y_{10} by an order of magnitude for most stabilities. This results in an impact area that is elliptical in shape (an ellipse is used here to approximate what is actually a "teardrop" shape) with a long major axis in the radial direction from the source and a relatively short minor axis in the crosswind direction. An example of an impact area an arbitrary distance from a source is illustrated in Figure 52. This configuration of the impact area under C_{max} conditions should be taken into account when proposing air monitoring sites.

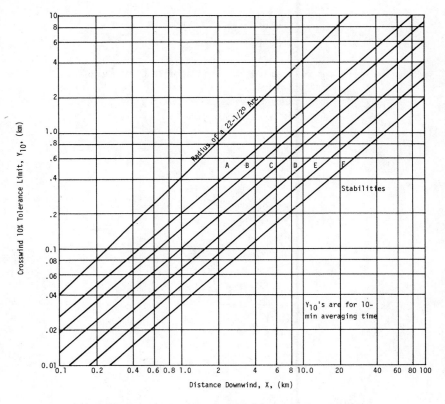

Figure 51. Y_{10} versus distance downwind for six stabilities.

The best method for determining the optimum number and location of monitoring sites is to take into account the size of the "representative area" under C_{max} conditions, the size of the potential zone requiring monitoring (from Figure 48), and the frequency of occurrence of meteorological conditions causing C_{max} values in each zone. A statistical approach is the most appropriate. The first step is to determine the ratio of the representative area of a monitoring station located in the center of a potential zone to the area encompassed by the zone (the zone is "donut" shaped if wind direction is not accounted for). If X_{10} < (radial dimension of the zone), then the proportion of the zone covered by a single monitoring station can be calculated from

$$CR_S = \frac{\text{Area of Ellipse}}{\text{Area of Sector}} = \frac{\frac{(X_{10}+Y_{10})^2}{4}}{(X_{Z_0}-X_{Z_1})^2} \qquad (3)$$

where CR_S = the coverage ratio for one station (without regard for wind direction)

X_{Z_1} = inside radius of circular zone, and

X_{Z_0} = outside radius of circular zone.

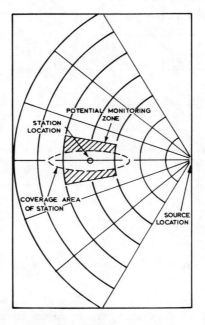

Figure 52. Illustration of the coverage area of a single station downwind of a point source (shaded area represents the area within the potential monitoring zone not covered by the station. The ratio of the coverage area to the total area of the potential zone equals the coverage ratio.).

Figure 52 depicts the representative area, a potential zone, and the coverage ratio, CR, graphically. Coverage ratios, CR, can be calculated for a single station, CR_S, or for all stations combined. The sum of all station coverage ratios equals the total coverage ratio, CR_T, for the network. To calculate coverage ratio for a single wind direction, multiply the CR from Equation 3 by the number of wind sectors into which the "donut-" shaped zone is divided.

The area where the maximum concentration, C_{max}, will occur divided by the area where it could occur is equal to the coverage ratio for a single station, and the probability, P, that one station located anywhere within the areas where C_{max} could occur will actually measure C_{max}. To calculate the probability that one station will measure the maximum concentration n times when the meteorological conditions causing C_{max} occur N times per year can be calculated using the binomial distribution function. [8]

$$P = \frac{N!}{n!(N-n)!} \, C.R.^n (1 \text{-} C.R.)^{N-n} \qquad (4)$$

For large values of N, the probability, P, or coverage ratio, CR, is best determined by using the normal approximation to the binomial distribution and solving for the statistic, Z. The probability associated with various Z values can be determined from the statistical table given in Table 7. Z is calculated using the equation

$$Z = \frac{n \text{-} NP}{(NP(1 \text{-} P))^{\frac{1}{2}}} \qquad (5)$$

The analysis of the Gaussian plume previously discussed indicates that the area where maximum ground level concentrations occur is very narrow in the Y-direction and quite long in the X-direction (see Figure 52). For this reason, the likelihood that a monitoring station will successfully observe maximum levels is highly dependent on its circumferential location about the source. For this reason, simplified estimates of CR for $X_{10} > (X_{z_0} \text{-} X_{z_1})$ are sufficient in most cases. CR is then determined by dividing $Y_{10}(x)$ by either the circumference of a circle, for all wind directions, or a single arc about the point source for one wind direction, having a radius equal to the downwind distance, X. The circumference can be calculated using the equation

$$\text{Circumference} = 2 \, \pi \, x \qquad (6)$$

Using this simplified method of estimating the coverage ratio of a station, CR_S, Figure 53 was prepared which gives coverage ratios for one station located within a single 22-1/2° wind sector plotted against downwind distance, x, for various stabilities. Using this figure, coverage ratios for single stations can be determined, assuming they are properly located in the downwind direction, x.

The calculated value for CR, plus the probable number of occurrences of the meteorological conditions producing C_{max} in a given zone can then be used to determine the probability of n successful measurements

Table 7.   Table of Probabilities Associated with Values as Extreme as Observed Values of z in the Normal Distribution[a]

z	0.00	0.01	0.02	0.03	0.04	0.05	0.06	0.07	0.08	0.09
0.0	0.5000	0.4960	0.4920	0.4880	0.4840	0.4801	0.4761	0.4721	0.4681	0.4641
0.1	0.4602	0.4562	0.4522	0.4483	0.4443	0.4404	0.4364	0.4325	0.4286	0.4247
0.2	0.4207	0.4168	0.4129	0.4090	0.4052	0.4013	0.3974	0.3936	0.3897	0.3859
0.3	0.3821	0.3783	0.3745	0.3707	0.3669	0.3632	0.3594	0.3557	0.3520	0.3483
0.4	0.3446	0.3409	0.3372	0.3336	0.3300	0.3264	0.3228	0.3192	0.3156	0.3121
0.5	0.3085	0.3050	0.3015	0.2981	0.2946	0.2912	0.2877	0.2843	0.2810	0.2776
0.6	0.2743	0.2709	0.2676	0.2643	0.2611	0.2578	0.2546	0.2514	0.2483	0.2451
0.7	0.2420	0.2389	0.2358	0.2327	0.2296	0.2266	0.2236	0.2206	0.2177	0.2148
0.8	0.2119	0.2090	0.2061	0.2033	0.2005	0.1977	0.1949	0.1922	0.1894	0.1867
0.9	0.1841	0.1814	0.1788	0.1762	0.1736	0.1711	0.1685	0.1660	0.1635	0.1611
1.0	0.1587	0.1562	0.1539	0.1515	0.1492	0.1469	0.1446	0.1423	0.1401	0.1379
1.1	0.1357	0.1335	0.1314	0.1292	0.1271	0.1251	0.1230	0.1210	0.1190	0.1170
1.2	0.1151	0.1131	0.1112	0.1093	0.1075	0.1056	0.1038	0.1020	0.1003	0.0985
1.3	0.0968	0.0951	0.0934	0.0918	0.0901	0.0885	0.0869	0.0853	0.0838	0.0823
1.4	0.0808	0.0793	0.0778	0.0764	0.0749	0.0735	0.0721	0.0708	0.0694	0.0681
1.5	0.0668	0.0655	0.0643	0.0630	0.0618	0.0606	0.0594	0.0582	0.0571	0.0559
1.6	0.0548	0.0537	0.0526	0.0516	0.0505	0.0495	0.0485	0.0475	0.0465	0.0455
1.7	0.0446	0.0436	0.0427	0.0418	0.0409	0.0401	0.0392	0.0384	0.0375	0.0367
1.8	0.0359	0.0351	0.0344	0.0336	0.0329	0.0322	0.0314	0.0307	0.0301	0.0294
1.9	0.0287	0.0281	0.0274	0.0268	0.0262	0.0256	0.0250	0.0244	0.0239	0.0233
2.0	0.0228	0.0222	0.0217	0.0212	0.0207	0.0202	0.0197	0.0192	0.0188	0.0183
2.1	0.0179	0.0174	0.0170	0.0166	0.0162	0.0158	0.0154	0.0150	0.0146	0.0143
2.2	0.0139	0.0136	0.0132	0.0129	0.0125	0.0122	0.0119	0.0116	0.0113	0.0110
2.3	0.0107	0.0104	0.0102	0.0099	0.0096	0.0094	0.0091	0.0089	0.0087	0.0084
2.4	0.0082	0.0080	0.0078	0.0075	0.0073	0.0071	0.0069	0.0068	0.0066	0.0064
2.5	0.0062	0.0060	0.0059	0.0057	0.0055	0.0054	0.0052	0.0051	0.0049	0.0048
2.6	0.0047	0.0045	0.0044	0.0043	0.0041	0.0040	0.0039	0.0038	0.0037	0.0036
2.7	0.0035	0.0034	0.0033	0.0032	0.0031	0.0030	0.0029	0.0028	0.0027	0.0026
2.8	0.0026	0.0025	0.0024	0.0023	0.0023	0.0022	0.0021	0.0021	0.0020	0.0019
2.9	0.0019	0.0018	0.0018	0.0017	0.0016	0.0016	0.0015	0.0015	0.0014	0.0014
3.0	0.0013	0.0013	0.0013	0.0012	0.0012	0.0011	0.0011	0.0011	0.0010	0.0010
3.1	0.0010	0.0009	0.0009	0.0009	0.0008	0.0008	0.0008	0.0008	0.0007	0.0007
3.2	0.0007									
3.3	0.0005									
3.4	0.0003									
3.5	0.00023									
3.6	0.00016									
3.7	0.00011									
3.8	0.00007									
3.9	0.00005									
4.0	0.00003									

[a]The body of the table gives one-tailed probabilities under H_O of z. The left-hand marginal column gives various values of z to one decimal place. The top row gives various values to the second decimal place. Thus, for example, the one-tailed p of $z \geqslant 0.11$ or $z \leqslant -0.11$ is p - 0.4562.

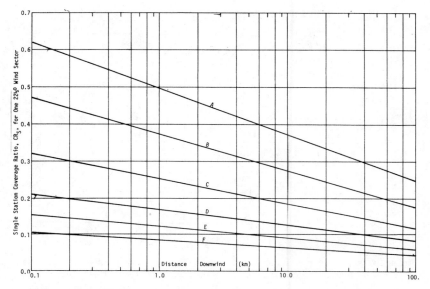

Figure 53. Single-station coverage ratios, CR_S, versus distance downwind for six stability classes (CR_S values are for 10 min averaging times).

of C_{max} using Equation 5; or if the number of desired successes is fixed, then the probability or coverage ratio required to assure at least n successes within a desired level of confidence can be calculated. General solutions of Equation 5 are given in Figure 54. This figure can be used to determine the total coverage ratio, CR_T, required to allow n successful measurements of C_{max} with 99% confidence given the frequency of occurrence, N, each year of the meteorological event producing C_{max}.

After solving the total coverage ratio required, the number of stations needed in the crosswind direction can be calculated.

$$N_S = \frac{CR_T}{CR_S} \tag{7}$$

where N_S =   number of stations needed
        CR_S =   coverage ratio of each station, and
        CR_T =   total coverage ratio required to achieve desired results.

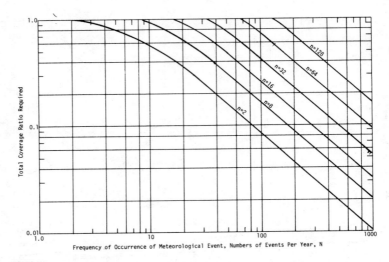

Figure 54. Coverage ratio required to observe C_{max}, n times, with 99% confidence, versus the number of occurrences, N, of meteorological phenomena causing C_{max}.

THE EFFECT OF AVERAGING TIME
ON COVERAGE RATIO

Coverage ratios are highly dependent on the extent of the representative sampling area in the Y-direction, which is in turn dependent on the averaging time of the sample. Several authors (*i.e.,* Hino,[9] Turner[5]) have pointed out the relationship between σ_y and sample averaging time, usually expressed as the exponential relationship

$$\sigma_y \propto (t)^{-0.5} \tag{8}$$

where t = averaging time in minutes (also see Figure 55). The physical cause of increasing σ_y with averaging time is due to the effects of meandering wind direction.

Because CR $\propto Y_{10} = 0.92\ \sigma_y(x)$, the coverage ratio varies with averaging time. The values of Y_{10} given in Figure 51 are for 10-min average concentrations. For longer averaging times the effect of meandering wind direction is to increase Y_{10}. Correction factors, CF, which can be used to increase Y_{10} and subsequently CR_S to account for averaging times different from 10 minutes can be calculated from the equation

$$CF = \frac{(10)^{-0.5}}{(t)^{-0.5}} = 0.316t^{0.5} \tag{9}$$

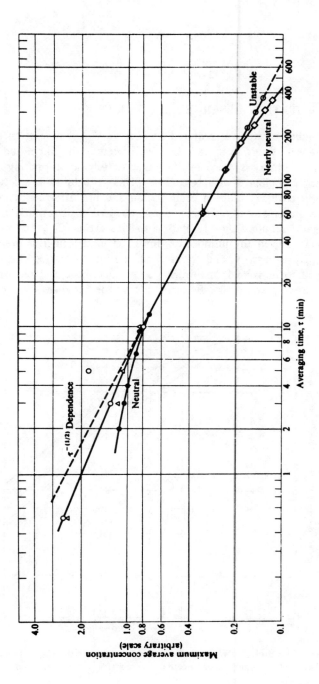

Figure 55. Dependence of the maximum concentration on averaging time (from Hino[9]).

Typical values are:
1-hr	2.4
3-hr	4.2
10-hr	7.7
24-hr	12.0

CAN ONE STATION OBSERVE C_{max} TWICE?

The number of stations required to observe C_{max} at least twice can be determined using Figure 54. However, what is the likelihood that C_{max} will be observed twice by the same station (appropriately applicable to checking compliance with NAAQS)? This probability is solely dependent on the coverage ratio of each station and not affected by the number of stations used. Figure 56 can be used to determine the coverage ratio required by one station to insure that C_{max} is observed at least twice, given the number of times the meteorological event causing C_{max} occurs, N, and the desired confidence level. The results can be used to indicate if monitoring some meteorological phenomena should be attempted at all. For example, if the coverage ratio of a

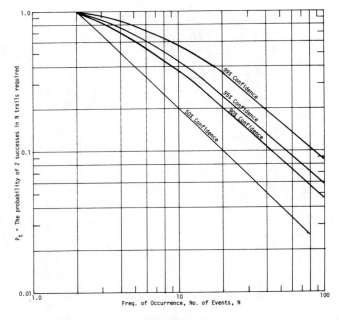

Figure 56. The probability required to successfully monitor C_{max} twice if the appropriate meteorological event occurs N times (for various confidence levels).

monitoring site is equal to 0.2 and the expected number of occurrences of meteorology causing C_{max}, N = 10, then the probability, or confidence level, that C_{max} will be measured at least twice is only 50%. Hence, there is only a 50:50 chance that any monitoring station will observe two occurrences of C_{max} even though it occurs 10 times. In such a case, it is probably inadvisable to attempt to monitor C_{max} under this condition at all. In light of the expense of air monitoring, sites should not be chosen to monitor C_{max} when the probability of success is low. Figure 56 shows how many times a meteorological event should occur in order to be 90% certain that C_{max} will be observed twice at one site for a range of values of CR_S. For example, for CR_S = 0.2 the meteorological event should occur at least 20 times each year to be 90% certain; for CR_S = 0.1 the same event must occur 43 times per year.

SELECTING SPECIFIC MONITORING SITES

Techniques for evaluating the number and location of air monitoring sites needed to monitor values of C_{max} have now been presented. The various techniques must now be integrated into a systematic approach to site selection. The steps are outlined below:

1. Evaluate historical meteorological data to determine prevailing meteorology and the frequency of occurrence of adverse conditions for air pollution dispersion. Choose ranges of meteorological conditions under which air monitoring may be desirable and directions from the source where conditions occur most frequently.

2. Using point source models (computer models, nomographs or general solutions presented herein), determine C_{max} and X_{max} under each range of meteorological conditions. From X_{max} identify potential zones where air monitoring stations may be needed.

3. Locate potential sites in the downwind directions where C_{max} values occur most frequently. Note that different C_{max} values occur under different types of meteorology, all of which may not occur most frequently in the same direction. Hence, this step does not simply mean to locate sites downwind of the prevailing wind direction.

4. Using Figures 48-56, determine the proper spacing and number of sites needed in the X direction to provide adequate coverage of each monitoring zone.

5. Determine the single station coverage ratio, CR_S, for the distances downwind and stability classes chosen in steps 1-3.

6. Calculate the total coverage ratio, CR_T, required by the network for each meteorological condition of interest and desired number

of observations, n. Divide CR_T by CR_S to determine number of stations needed in the crosswind direction.

7. Finally, calculate the probability of successful observations of C_{max} for each station, and if this probability is very low then some stations may be eliminated.

It is important to remember that the figures given are based on a tolerance range for C_{max} of 10%. If a different tolerance range is desired, then new curves should be prepared from the basic equations presented in the text. The steps in the site selection process remain the same.

The site selection procedure presented herein is designed for locating sites having the highest probability of successfully measuring maximum ground level pollutant concentrations occurring near point sources. In practice other objectives may be equally important, such as monitoring pollutant levels near sensitive receptors. Therefore, the sites identified using the procedures presented here should also be evaluated to see if certain site locations can be altered in order that multiple objectives might be served by the same site or if additional sites are needed.

REFERENCES

1. "Selective Guide to Climatic Data Sources," Environmental Data Service, U.S. Dept. of Commerce, ESSA, Pub. No. KMRD, No. 411.
2. Briggs, G. A. "Plume Rise," U.S. AEC Division of Information (1969).
3. Fay, J. A., et al. "A Correlation of Field Observations of Plume Rise," J. Air Pollution Control Assoc. 20:391-397 (1970).
4. Montgomery, T. L., et al. "A Simplified Technique Used to Evaluate Atmospheric Dispersion of Emissions from Large Power Plants," J. Air Pollution Control Assoc. 23(5):388-394 (May 1973).
5. Turner, D. B. "Workbook of Atmospheric Dispersion Estimates," U.S. EPA Pub. No. AP-26 (revised 1970).
6. Pasquill, F. Atmospheric Diffusion (New York: Van Nostrand, 1969).
7. STAR Program: Seasonal and Annual Wind Distribution by Stability Class (6), National Climatic Center, Federal Building, Ashville, North Carolina.
8. Miller, I. and J. Freund. Probability and Statistics for Engineers (Englewood, New Jersey: Prentice-Hall, Inc., 1965).
9. Hino, M. "Evaluation of Point Source Data," Atmos. Environ. 2:149 (1968).

CHAPTER VIII

AIR SAMPLE COLLECTION

COLLECTION OF AIR SAMPLES

Introduction

This section discusses the handling of air samples before analysis for pollutant concentration. The design of air sampling equipment depends on the sample size required by the analyzer and the retention time of the sample before analysis. Limitations on retention time depend on the reactivity of the pollutant being measured. While pollutants such as photochemical oxidants and nitrogen dioxide require very short time periods between sample collection and analysis, more inert pollutants, such as carbon monoxide samples, may be stored in special sampling bags for hours, or even days, before analysis. Particulate matter, collected on filter paper, is routinely analyzed several days after the sample is taken. Several methods available for collecting the sample depend on: (a) the air pollutant to be analyzed, (b) the analytical technique employed, the number of samples desired, and (c) the averaging time of the pollutant concentration measurement.

Three methods of air sampling that find common application in air monitoring are continuous sampling, intermittent sequential sampling and integrated grab sampling. Continuous sampling of air pollution produces a continuous profile of the air pollution concentration governed only by the response time of the analyzing instrument. This provides information on short-term fluctuations in pollutant level. This is illustrated in the hypothetical air pollution measurement shown in Figure 57a.

Intermittent sequential sampling (ISS) involves one analyzer and air samples from more than one sampling probe (connected at a central sampling manifold). The air pollution analyzer measures an air sample from one probe at a time depending on which manifold valve is open (see Figure 57b).

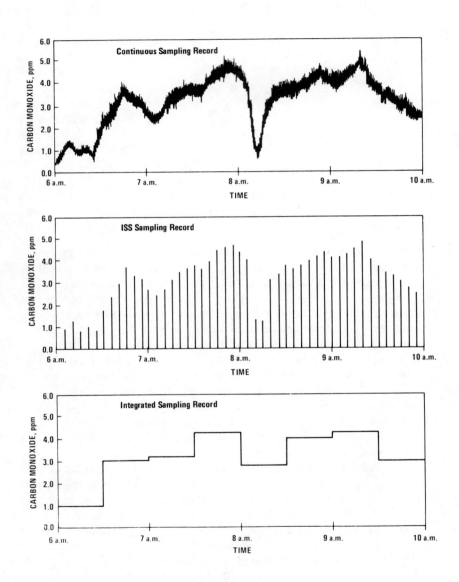

Figure 57. (a) Continuous sampling record, (b) ISS sampling record, (c) integrated sampling record.

As a result, samples are taken from each probe at a frequency equal to the cycle-time of the manifold, *i.e.,* the response time of the analyzer multiplied by the number of probes. In the case of a carbon monoxide analyzer operating at a sample response time of 30 seconds, and with five probes in the sampling network, the cycle-time would be 2-1/2 min. Thus, 24 measurements could be made each hour at each probe.

The advantage of intermittent sequential sampling is that it provides measurements from several sampling probes while, like continuous sampling, identifying many of the fluctuations in air pollution concentrations that occur during a single one-hour period. However, the ISS method will not reproduce the same concentration profile as a continuous analyzer operating from a single probe because some of the instantaneous peaks in pollution concentration will be lost (see Figure 57b).

The third type of air sampling commonly used is the integrated grab sample, in which the air sample is continuously collected in a plastic bag, in a liquid absorbing reagent or on a filter paper. The air flow is maintained at a constant rate for the duration of the sample, typically 15 min, 1 hr, or 24 hr and is then turned off. The sample may be stored or analyzed immediately. The significant difference between this type of sampling and the two previously discussed is that grab sampling provides only a single concentration measurement, with an averaging time that is equal to the duration of the sample. Figure 57c illustrates how the same pollutant concentration pattern shown in Figure 57a would appear if sampled for 30-min intervals using an integrating grab sampling method. The advantage of grab sampling is that samples can be taken from many locations simultaneously and analyzed afterward.

CONTINUOUS AIR MONITORING

Introduction

Continuous air monitoring requires a considerable amount of supportive equipment to collect the air sample from the atmosphere and transport it to the analyzer without altering the original air pollution concentrations. The types of supportive equipment needed to perform continuous monitoring include sampling probes, sample conditioning equipment (filters, driers, humidifiers, temperature controllers, etc.), analyzers, recorders, calibration and maintenance equipment, and shelters (see Figure 58).

Sampling Probes and Manifolds

Most continuous air pollution monitoring instruments are housed in either permanent or portable shelters to provide protection from the

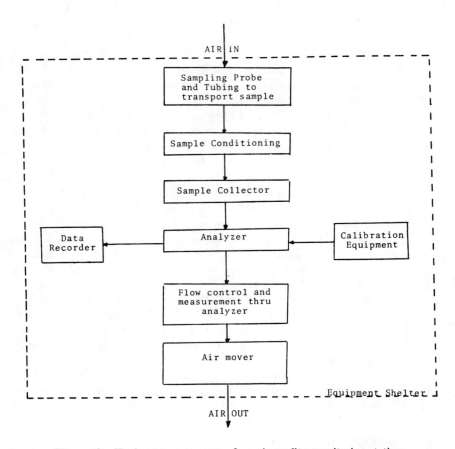

Figure 58. Hardware components of an air quality monitoring station.

weather. The air sample to be analyzed is then transported from outside the building to the instrument. This is usually accomplished using a sampling manifold similar to that pictured in Figure 59. Since the sampling manifold projects above the building roof with the sample inlet at the top, the manifold inlet is usually capped or so shaped that rainfall cannot enter.

Each continuous air monitoring instrument is provided an access port to the sampling manifold. When a glass sampling manifold is used, these sampling ports are prefabricated. When the sampling manifold is constructed of plastic or stainless steel pipe, sampling ports can be drilled through the wall of the pipe. Sample inlet lines to the analyzers usually begin at

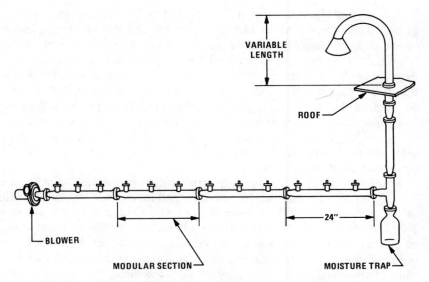

Figure 59. Sampling manifold for continuous air monitoring.

a point near the center of the manifold. This helps reduce interference due to dirty walls inside the sampling manifold.

An ample supply of ambient air is usually pulled through the manifold using a centrifugal blower or other sampling pump. The sampling pump controls the air volume exchange to 300 to 1000 times per minute, providing a retention time inside the manifold of 0.5 to 2 sec. This is to insure against sample degradation before analysis. A recent experiment reported by Decker et al.[1] indicated that when a 1-in. (0.4 cm) o.d. glass manifold 9 m long was operated at a flow rate of approximately 3 cfm, no measurable losses in ozone concentration in the sample air were observed. It is unlikely that other pollutants of interest, such as SO_2, NO, NO_2, THC, NMHC and CO are modified due to wall effects in the air sampling system if ozone losses are small.

The rate at which the sample enters the instrument varies from analyzer to analyzer but usually ranges from 0.1 to 10 liters per minute. The inlet air sample is usually pulled through a short length (less than 10 ft) of small diameter inert tubing, the diameter of which is fixed according to the diameter of the inlet opening of the analyzer. The sample lines are kept as short as possible in order to minimize the retention time of the sample in the line before analysis, thus decreasing the likelihood of alterations in the original pollutant concentration due to adsorption, desorption, or chemical reactions within the sample line.

The physical dimensions of the sampling manifold depend to a great extent to the size of the air monitoring shelter and the number of instruments that will be connected to the manifold. A survey, conducted by Yamada,[2] of 34 air pollution control agencies operating continuous air monitoring stations during 1968 indicated that the typical sampling manifold had a diameter of 1/2-in., was 14 ft long, and extended 5 ft above the roof of the station. Sample flow rates typically ranges from 2 to 12 liters/min. Elfers[3] suggests a much larger diameter tubing (50 mm or greater i.d.) with an air flow rate between 3 and 5 cfm (accomplished using a small blower rated at 60 cfm) since smaller dimensions are not optimal in most applications.

Sample interference is frequently attributed to a chemical or physical reaction either with particulate materials that have coated the inside walls of the manifold or with the material from which the manifold is constructed. Frequent cleaning of the manifold can eliminate particulates as a source of interference while the second source of trouble can be avoided by proper choice of construction material. According to Elfers, who sites three studies that have been conducted to determine the suitability of materials, such as polypropylene, polyethylene, PVC, Tygon, aluminum, brass, stainless steel, copper, Pyrex glass and Teflon for use as intake sampling lines, only Pyrex glass and Teflon are acceptable for all continuous monitoring equipment. At some point in the sample line, between the sampling manifold and the analyzer, a small diameter (1-2 in.) inline filter is often used to protect the analyzer from clogging and chemical interference due to particulate buildup in the sample lines or analyzer inlet. These filters can be of the Whatman type and should be changed every few days of continuous operation, depending on ambient conditions.

Shelter Design

To insure against equipment damage, continuous air monitoring equipment must be housed in protective shelters, which may range from small metal or wooden structures protecting against rain damage to permanent buildings containing monitoring instruments plus wet laboratory facilities. Portable buildings of mobile home type structure, which are available in a wide range of sizes with prefabricated utility connections, are frequently used. These buildings cost from $10 to $40 per square foot and can be moved to new sampling sites when necessary. The size of the shelter required depends on the purpose for which it is intended and the quantity of air monitoring equipment to be operated in the station. Bryan[4] recommends 35 sq ft of floor space for each analyzer, plus additional space as required for desk space, gas bottle storage, toilet facilities, etc.

Available space can be utilized most efficiently when instruments are rack mounted rather than lying on tables or work benches.

Continuous air quality monitoring instruments are delicate and should be protected from vibrations, direct sunlight and fluctuating temperatures inside the shelter. Instruments can be isolated from vibrations by using rubber shock mounts or other effective methods. The shelter should be heated in the winter and air conditioned in the summer to maintain an air temperature near 70°F.

INTERMITTENT SEQUENTIAL SAMPLING

Application

Intermittent sequential sampling can be useful for determining the microscale air pollution concentration distribution within about 1000 ft of a transportation corridor in order to validate and/or calibrate microscale simulation models. Intermittent sequential sampling requires a single analyzer, a specially designed sampling manifold and a network of sampling probes that can be located at various horizontal and vertical distances from the pollution source to be monitored (see Figure 60).

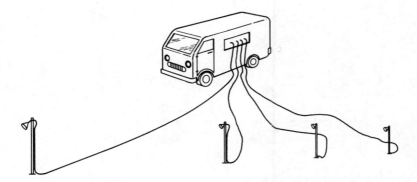

Figure 60. Pictorial of an intermittent sequential sampling field operation.

Sample Tubing and Manifold Requirements

Intermittent sequential sampling can require large amounts of tubing depending on the number of probes, the distance spanned between probes and the location of the sample manifold and analyzer. This tubing should be small diameter (1/8 to 3/8 in.) plastic tubing constructed of material that will not affect the air sample before analysis. Noll *et al.*[5] have reported that air samples drawn through tygon tubing exposed to direct sunlight resulted in a false positive response from a nondispersive infrared carbon monoxide analyzer. Such interference can be eliminated by covering the tubing with aluminum foil or by placing the tubing in plastic or metal electrical conduit.

The ISS manifold is the heart of this type of monitoring system. A simple manifold designed for four sampling points is illustrated in Figure 61. The basic function of the manifold is to provide a branch through which a single analyzer can draw air samples from a number of sampling probes. The air flow through any passageway in the branch is controlled by using a stopcock or other type of valve. When all valves except one

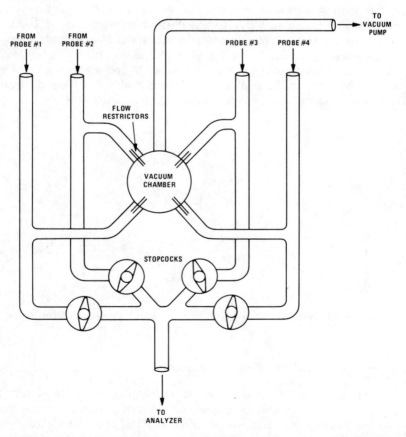

Figure 61. Schematic diagram of an intermittent sequential sampling manifold.

are closed, the air sample will be drawn through a single probe line. By opening and closing each valve alternately, air samples can be taken from each probe in a sequential pattern. After each probe is sampled, the cycle is repeated.

An important component of the ISS manifold is the purge flow system. Just upstream of each sampling valve is a "T" in the sample line, one side of which is connected to a vacuum chamber which causes an air sample to be drawn continuously through each sample line. The purpose of this air flow is to purge lines between samples when the sampling valve is closed. The rate of air flow through each sample line is controlled by using a critical orifice, needle valve or other flow-limiting device. The purge air flow rate is an important parameter in determining the averaging time of the sample that is analyzed.

ISS manifolds can be designed to be operated manually or electronically. Manually operated manifolds can be built very cheaply using glass or plastic tubing, stopcocks, a glass bottle vacuum chamber, rubber stoppers and critical orifices made from glass pipettes or hypodermic needles. More sophisticated ISS manifolds employ computer controller, electrically activated solenoid valves and individually metered purge flow lines. The bulk of the expense of the more sophisticated systems lies in the costs of solenoid valves, flow meters, fabrication and the computer hardware and software required to control the system automatically.

Shelter Design

As in the case of continuous monitoring, the analyzer used for intermittent sequential sampling must be sheltered from the elements. Since most air monitoring investigations using ISS are relatively short-term studies, use of a permanent type structure for housing the air pollution analyzer and sampling manifold is usually impractical. For this reason, it is usually advantageous to install the analyzer and manifold in a mobile vehicle such as a camping trailer, a mobile van or a station wagon.

Design Limitations

In general, intermittent sequential sampling should be limited to 5 to 10 sampling probes per analyzer to keep the cycle-time on the order of several minutes. With a short cycle-time, the ambient pollution levels at each probe can be monitored many times each hour, providing a fairly representative picture of the profile of air pollution concentrations that occur during that period. Because of the required delay between initial sample collection and analysis inherent in all types of intermittent sequential sampling, it is generally inapplicable for use in sampling the reactive air pollutants such as photochemical oxidants, reactive hydrocarbons and nitrogen dioxide. It can best be used for sampling carbon monoxide and nitric oxide (NO), which are the least reactive of the important air pollutants.

Intermittent sequential sampling can generate from 60 to over 200 discrete air pollution concentration measurements every hour. This represents a sizable quantity of air quality data and requires that adequate measures be taken to collect and record each measurement. This can be accomplished using strip chart recorders or electronic data loggers. Strip chart recorders will record the continuous output of the analyzer, thus recording all of the measurements, but it is difficult to separate the different measurements from each probe. This is possible if a high-speed chart drive is used (1-5 cm/min) and probe identification markings are written on the chart paper periodically. Electronic data loggers are most applicable for sophisticated computer-controlled ISS systems. When data loggers are used, it is important that the data logger scan the analyzer at precisely the right time to record the analyzer output, which is equivalent to the pollutant concentration at the probe being sampled. This requires that the analyzer, manifold and data logger all be precisely synchronized, providing the proper lag time between the initiation of the sample by the manifold and the full response of the analyzer. Manual reduction of data is also possible. A single efficient operator can use a manual ISS manifold and a continuous analyzer, recording each sample measurement manually on specially prepared data forms. To reduce the total quantity of manual work required, the data form can be designed as a computer key punch sheet to be used with a standard digital computer.

INTEGRATED GRAB SAMPLING

Application

Because the sampling equipment required is usually portable and less expensive than continuous analyzers, integrated grab sampling is the most flexible of the air sampling methods discussed in this chapter. Grab sampling programs can be designed to study microscale and mesoscale air pollution concentration distributions. Several air samplers can be operated in close proximity to each other, as in a microscale study, or they can be located at various points throughout an urban area, as in a mesoscale study. After collection, each grab sample must be analyzed in the laboratory for the concentration of air pollution.

Particulate Samplers

The most commonly used samplers for grab sampling ambient concentrations of particulate matter are the high volume sampler (hivol) and the tape sampler, both of which use a vacuum pump to pull an air sample through filter paper, which removes the particulates from the air stream. The high volume sampler uses a high air flow rate (50 cfm) and collects particles on an 8-1/2 in. x 11 in. filter paper. The duration of the sampling period is standardized at 24 hr. The tape sampler, on the other hand, uses a much lower flow rate (0.25 cfm) and collects particulate matter on a continuous 1-1/2 in. wide filter tape. The tape sampler is usually adjusted to take a 2-hr average air sample.

Bubbler Sampler

Integrated grab sampling for gaseous pollutants commonly utilize a bubbler sampler, which uses a vacuum pump to pull a regulated air flow through a midget impinger or fritted bubbler containing a liquid-absorbing reagent. After a sampling period of from 1 to 24 hr, the sampling pump is turned off and the liquid reagent can be collected and taken to a laboratory for analysis.

The types of gaseous air pollutants that can be measured using bubbler sampling methods are numerous: nitric oxide, nitrogen dioxide, total nitrogen oxides, photochemical oxidants and various atmospheric hydrocarbons. The sampling apparatus is basically the same regardless of the type of air pollutant being measured; however, the chemical composition of the liquid absorbing reagent is different for each different air pollutant. The *Federal Register*[6] and analytical texts such as *Methods of Air Sampling and Analysis*[7] describe in detail how to prepare absorbing reagents for the collection and analysis of many different gaseous air pollutants.

Bag Sampling

Another method for grab sampling gaseous air pollutant concentrations is to collect and store an air sample in a volumetric container. Rigid containers can be used, but a much more convenient method is to use plastic bags. A low-capacity diaphragm pump is used to fill the plastic sampling bag slowly over a period of 15 min to 1 hr. The volumetric capacity of the sampling bag and the duration of the sample will determine the air flow rate that is required to fill the bag. At the end of the sampling period, the pump is turned off and the bag sample removed. Each sample bag has an adjustable valve that can be closed after the bag has been filled in order to prevent any leakage of the air sample.

Examples of the types of materials used in the construction of plastic air sampling bags are Aluminized Scotch-pak, Mylar, Teflon, Tedlar and Saran. Since some air pollutants will react chemically with these, it is desirable to run "decay" curves on a number of samples in order to establish a reasonable storage time before analysis. Certain air contaminants, such as ozone, cannot be sampled or stored in plastic bags at all because of their high chemical reactivity. Carbon monoxide is the pollutant most appropriate for collection in this manner. According to Pinkerman,[8] 3-mil Tedlar bags can be used with reasonable confidence for the sampling of ambient levels of total nitrogen oxides (NO_x). However, the bags cannot be used to determine the concentrations of either nitric oxide (NO) or nitrogen dioxide (NO_2) because NO-NO_2 transformations may occur while the sample is being stored.

Another important consideration when choosing the proper air sampling bag is the long-term life of the bag and its flexibility. Aluminized Scotch-pak and Tedlar are both flexible materials which can be used as many as 50 times. Teflon bags become brittle with time and flexing causes the bags to split at the seams.

The volume of the sampling bag must provide a sufficient air sample for analyses. In some cases, it is desirable to divide the contents of a single bag among several analyzers in order to measure the concentration of several different air pollutants in the same sample. The total volume of air sample required is then equal to the sum of the air sample volumes required by each analyzer. A five-liter sampling bag will normally provide an adequate air sample for several analyzers. In general, the total capacity of the sampling bag should be about twice the total volume of sample needed for all analyzers.

Sample bags can be filled by using a small diaphragm pump mounted inside an air sampling box designed to contain the sample pump, a battery to provide power for the pump, a timing mechanism to turn the pump on and off, and sufficient space for the sample bag to be filled (see Figure 62). The sample box can include a sample probe with the inlet located at a height of 5 ft above the ground (the breathing zone).

Manpower Required and Data Recording

A great disadvantage of all grab-type air sampling is the large number of man-hours required to collect the samples and analyze each sample. Many air monitoring investigations employing grab sampling methods require the services of two to three full-time technicians as well as a station wagon or truck for collecting the samples. The total number of man-hours and miles driven each day depends, to a large extent, on the number of

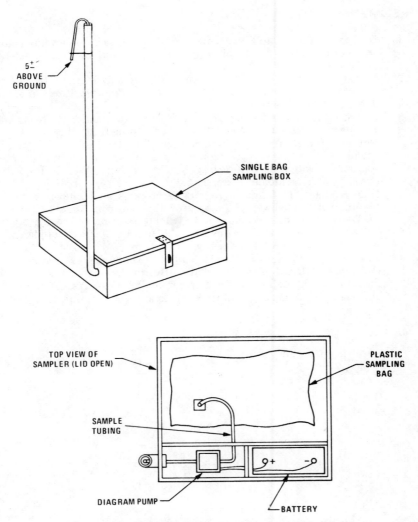

Figure 62. Pictorial illustration of a single bag air sampling box.

times each station must be visited. For example, air sample boxes that can operate unattended for 12 hr a day may have to be visited only once a day, whereas sample boxes which take only one sample per hour will have to be visited 10 to 12 times per day in order to collect an entire day's data.

REFERENCES

1. Decker, C. E., T. M. Royal and J. B. Tommerdahl. "Development and Testing of an Air Monitoring System," Contract No. 68-02-1011 (1973).
2. Yamada, V. M. "Current Practices in Siting and Physical Design of Continuous Air Monitoring Stations," *J. Air Pollution Control Assoc.* 20(4):209-213 (April 1970).
3. Elfers, L. A. "Field Operations Guide for Automatic Air Monitoring Equipment," Contract No. CPA-70-124 (Springfield, Virginia: National Technical Information Service, July 1971).
4. Bryan, R. J. "Air Quality Monitoring," in *Air Pollution, Vol. II, Analysis, Monitoring, and Surveying*, Arthur Stern, Ed., 2nd ed. (New York: Academic Press, 1968).
5. Noll, K. E. *et al.* "Annual Report of Progress on Monitoring Air Pollution near Highways," submitted to Tennessee State Department of Transportation, Environmental Planning Division (1974).
6. "National Primary and Secondary Ambient Air Quality Standards," *Federal Register* 36(84) (April 30, 1971).
7. Intersociety Committee. *Methods of Air Sampling and Analysis* (Washington, D.C.: American Public Health Association, 1972).
8. Pinkerman, K. California Department of Public Works, Division of Highways, Sacramento, California. personal conversation (July 1974).

CHAPTER IX

ANALYTICAL METHODS
FOR MEASURING AIR POLLUTANTS

INTRODUCTION

This chapter presents analytical methods which can be used for quantitative analysis of the concentration of pollution in a sample of ambient air. The purpose of the chapter is to present an overview of the common analytical methods used in air pollution monitoring. EPA has established reference methods for the measurement of the following criteria air pollutants: carbon monoxide (CO), nitrogen dioxide (NO_2), nonmethane hydrocarbons (NMHC), photochemical oxidants (specific for ozone, O_3), sulfur dioxide (SO_2) and total suspended particulates (TSP). Although these reference methods of measurement are specific for the pollutants for which they are intended, many commercially available analytical methods are *not* specific for criteria pollutants and frequently measure different pollutant species. For example, the federal reference method for nonmethane hydrocarbons requires a gas chromatograph and a flame ionization detector; however, the flame ionization instrument is frequently used alone to measure "total hydrocarbons" (methane plus other hydrocarbons). Another federal reference method measures nitrogen dioxide. However, continuous instruments are frequently used that measure nitrogen dioxide plus nitric oxide (commonly called "total nitrogen oxides, NO_x"). Another federal reference method is for photochemical oxidants measured using the chemiluminescent method which is specific for ozone (O_3); however, many instruments are frequently used that measure "total oxidants," *i.e.*, ozone plus other oxidizing agents found in polluted atmospheres. Such methods, which are not specific for criteria pollutants, do not produce air quality data that is directly comparable to National Ambient Air Quality Standards. The problem is further complicated because some states, such as California, have air quality standards that are based on different reference methods of measurement than the federal standards. In such cases, different or additional analytical methods may be required to check for compliance with air quality standards.

CATEGORIES OF ANALYTICAL METHODS

Methods for measuring air pollutants fall into one of three categories: (a) approved, (b) unacceptable, and (c) those methods that are neither approved nor unacceptable (unapproved).[1] The official methods are the federal reference methods described in the appendices to 40 CFR Part 50, originally promulgated on April 30, 1971 (36FR8186) with the National Ambient Air Quality Standards (NAAQS). This *Federal Register* also introduced the concept of an "equivalent method," which is any method that can be demonstrated to be "equivalent" to the reference method. Thus, unapproved methods may become approved only by demonstrating equivalence to the reference method. Table 8 lists those analytical methods that have been designated as "approved," "unapproved," or "unacceptable."

Table 8. Designation of Analytical Methods for Air Pollution Measurement[1,2]

Pollutant Code		Method	Approved	Unapproved	Unacceptable
TSP 11101	91	Hi-Vol (FRM)[a]	X		
CO 42101	11	NDIR (FRM)	X		
	12	Coulometric		X	X
	21	Flame ionization		X	
NO$_2$ 42602	11	Colorimetric		X	
	12	Colorimetric		X	
	13	Coulometric		X	
	14	Chemiluminescence		X	
	71	J-H bubbler (orifice)			X
	72	Saltzman			X
	84	Sodium arsenite (orifice)		X	
	91	J-H bubbler (frit)			X
	94	Sodium arsenite (frit)		X	
	95	TEA		X	
	96	TGS		X	
Photochemical					
O$_x$ 44101	11	Alkaline KI instrumental			X
(Ozone)	13	Coulometric			X
	14	Neut KI colorimetric		X	
	15	Coulometric		X	
	51	Phenolphthalin			X
	81	Alkaline KI bubbler			X
	82	Ferrous oxidation			X
44201	11	Chemiluminescence (FRM)	X		
Nonmethane Hydrocarbons					
NMHC 43102	11	Gas chromatograph-flame ionization	X		

[a]FRM = Federal Reference Method

Regulations require that a method be tested according to prescribed procedures and meet certain prescribed specifications to be approved as an equivalent method. In essence, manual methods must demonstrate a consistent relationship to the reference method in side-by-side measurements of ambient air. Automated methods must also meet certain performance specifications.

GENERAL ANALYTICAL METHODS

Measuring the concentration of air pollution in a sample of ambient air is an application of quantitative analytical chemistry. Methods of analysis rely on the detection of chemical reactions and/or physical properties that are associated with the presence of the air pollutant being measured. Although a detailed treatment of analytical methods of air pollution analysis is beyond the scope of this chapter, a basic understanding of the most common methods used for field monitoring of air pollutants is important.[2,3] The analytical methods that will be described include colorimetry, coulometry, chemiluminescence, IR and UV absorption spectroscopy, flame ionization, gas chromatography and gravimetry.

Colorimetry

Colorimetry is defined as an analysis in which the quantity of a colored substance is determined by measuring the relative amount of light passing through a solution of that substance. The constituent may itself be colored and thus be determined directly, or it can be reacted with a reagent to form a colored compound and thus be determined indirectly. The physical law that underlies colorimetric analysis is commonly referred to as the Beer's law, which states that the degree of light absorption by a colored solution is a function of the concentration and the length of the light path through the solution. Colorimetric analysis is commonly employed for the continuous measurement of three air pollutants: sulfur dioxide, nitrogen dioxide and oxidants.

Figure 63 shows a typical colorimetric monitor. An atmospheric sample is drawn into the air-reagent flow system, first entering the absorber or scrubber where the desired constituent is reacted with the appropriate reagent. The air sample is separated from the reacted reagent and passed through the air pump and finally discharged to the atmosphere. The reacted reagent passes from the absorber into the continuous flow colorimeter where the absorbance is measured. In most cases a dual flow colorimeter, as shown, is employed for measuring the absorbance of the unreacted reagent initially.

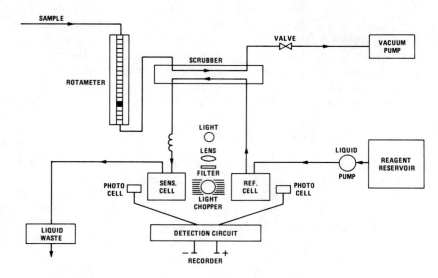

Figure 63. Schematic diagram of a typical colorimetric monitor.

Coulometry

Coulometry is a mode of analysis in which the quantity of electrons required to oxidize or reduce a desired substance is measured. This measured quantity, expressed as coulombs, is proportional to the mass of the reacted material according to Faraday's Law. Coulometric titration cells for the continuous measurement of sulfur dioxide, oxidants and nitrogen dioxide have been developed. Figure 64 is a diagram of a coulometric monitor.

One type of continuous air monitor is designed to respond to materials that are oxidized or reduced by halogens and/or halides. Upon introduction of a reactive material, the halogen-halide equilibrium is shifted. The system is returned to equilibrium by means of a third electrode which regenerates the depleted species. The current required for this generation is measured and is directly proportional to the concentration of the depleted species. This mode of analysis can be classified as secondary coulometry, usually employing a dynamic iodimetric or bromimetric titration. Most commercial coulometric systems are made specific by the use of prefiltration devices, scrubbers and/or chromatographic techniques that retain interfering compounds and permit passage of the desired constituents.

Another type of coulometric cell, designed for oxidant analysis, employs amperometry. The oxidant reacts with an iodine solution within

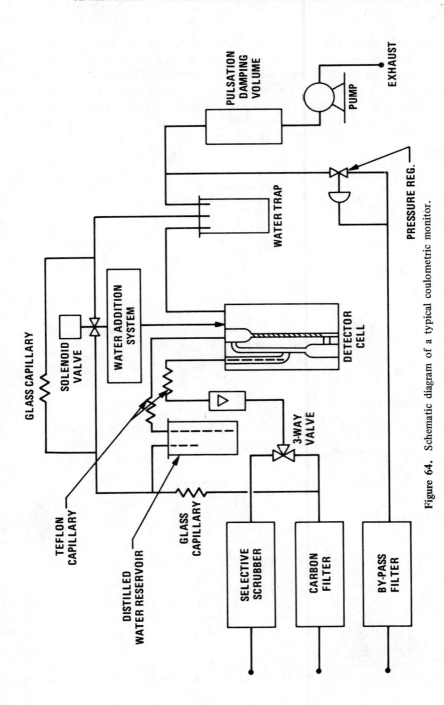

Figure 64. Schematic diagram of a typical coulometric monitor.

the cell, releasing iodine which depolarizes the cathode, thus permitting current flow which is proportional to the oxidant concentration. By passing reagent and sample over the electrodes, a continuous measurement is achieved. This type of monitor is also subject to interferences. Materials that undergo oxidation will appear as negative interferences; those that undergo reduction will appear as positive interferences. Figure 65 is a diagram of a typical amperometric-type monitor.

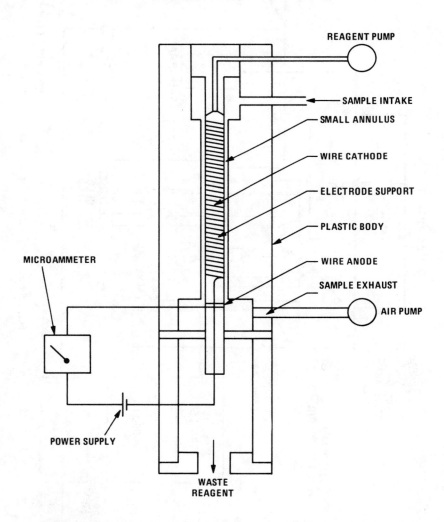

Figure 65. Schematic diagram of a typical amperometric monitor.

Chemiluminescence

Chemiluminescent detection techniques are employed for the measurement of atmospheric ozone and oxides of nitrogen. One method for measuring ozone employs the reaction of Rhodamine B impregnated on activated silica gel with ozone; the chemiluminescence produced is detected with a photomultiplier tube. Ambient air to be measured and ethylene are delivered simultaneously to a mixing cell where ozone reacts with the ethylene to emit light which is measured by a photomultiplier tube. If the air and ethylene flow rates are constant, the resulting photomultiplier signal can be related to the input ozone concentration. Figure 66 is a diagram of a gas-phase ozone monitor. The chemiluminescence that occurs when NO and O_3 react chemically provides a method by which nitric oxide can be measured. Chemiluminescent analyzers that can measure ambient levels of both NO and NO_2 are available.

IR and UV Absorption Spectroscopy

Infrared or ultraviolet radiation is passed through a chamber containing the air sample, after which the radiation is dispersed and detected. Because each chemical compound in the air sample absorbs the radiation in a characteristic pattern, the amount of radiation absorbed at various wavelengths can be used to identify the chemical component. The amount of light absorbed is proportional to the concentration of the component; therefore, a quantitative analysis may be obtained. An example of an infrared absorption spectrometer is the NDIR (nondispersive infrared) analyzer used for measuring carbon monoxide. A typical analyzer (Figure 67) consists of a sampling system, two infrared sources, sample and reference gas cells, detector, control unit and amplifier and recorder. The reference cell contains a noninfrared-absorbing gas while the sample cell is flushed with the sample atmosphere. The detector consists of a two-compartment gas cell (both filled with carbon monoxide under pressure) separated by a diaphragm whose movement causes a change of electrical capacitance in an external circuit and ultimately an amplified electrical signal.

During operation, an optical chopper intermittently exposes the reference and sample cells to the infrared sources. A constant amount of infrared energy passes through the reference cell to one compartment of the detector cell while a varying amount of infrared energy, inversely proportional to the carbon monoxide concentration in the sample cell, reaches the other detector cell compartment. These unequal amounts of

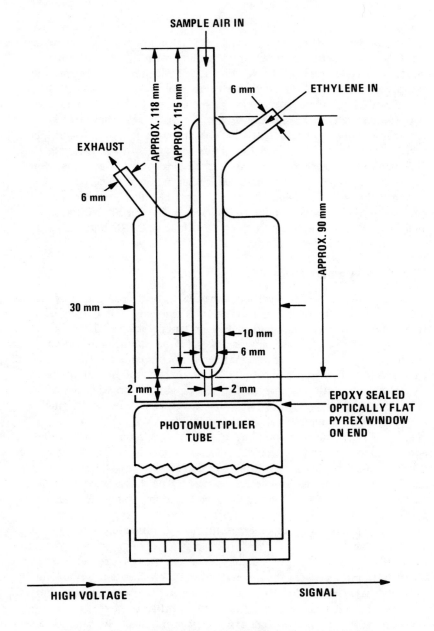

Figure 66.   Schematic diagram of a typical gas-phase chemiluminescent ozone detector.

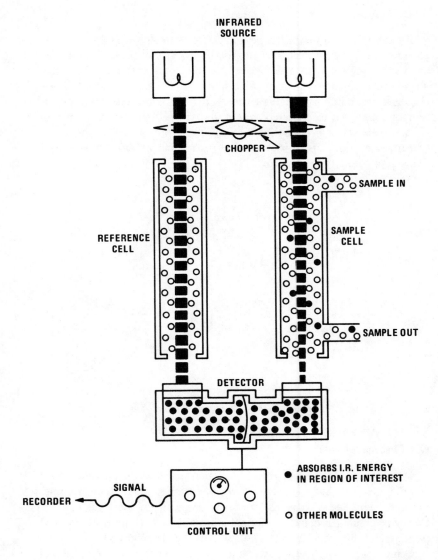

Figure 67. Schematic diagram of a typical nondispersive infrared CO monitor.

residual infrared energy reaching the two compartments of the detector cell cause unequal expansion of the detector gas, resulting in variation in the detector cell diaphragm movement, which produces the electrical signal.

Flame Ionization

Instruments employing flame ionization detectors have had wide application as a continuous monitor for hydrocarbons. The sample to be analyzed is mixed with a hydrogen fuel and passed through a small jet; air supplied to the annular space around the jet supports combustion. Carbon-containing compounds carried into the flame result in the formation of ions. An electrical potential across the flame jet and an ion collector electrode produce an ion current proportional to the number of carbon atoms. Figure 68 is a schematic diagram of a typical hydrogen flame ionization type monitor.

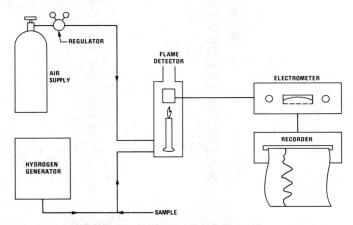

Figure 68. Schematic diagram of a typical flame ionization monitor.

Gas Chromatography

Chromatography is a technique for the separation of closely related compounds (*i.e.*, hydrocarbons). In gas chromatography, the sample is vaporized and the mixture is passed by a stream of inert gas (carrier) through a rigid container (column) containing a packing material. The packing has different affinities for each particular component in the mixture, and each component passes through at a different rate. As each component emerges from the column, it is observed by a sensitive detecting device (commonly a flame ionization detector). A diagram of a basic gas chromatograph is shown in Figure 69.

Investigations involving automated gas chromatographic separation have led to the development of a semi-continuous monitor for CO, CH_4 and total hydrocarbons. Measured volumes of air are delivered to a hydrogen

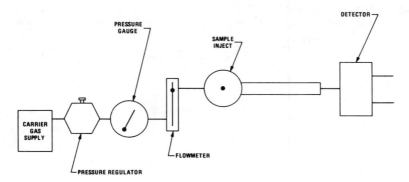

Figure 69. Basic diagram of a gas chromatograph (single column).

flame ionization detector to measure total hydrocarbon (THC) content. The same air sample is introduced into a stripper column which removes water, carbon dioxide and hydrocarbons other than methane. Methane and carbon monoxide are passed gravitatively to a gas chromatographic column where they are separated. The methane is eluted first and passes through a catalytic reduction tube into the ionization detector. Then carbon monixide is eluted into the catalytic reduction tube where it is reduced to methane before passing through the flame ionization detector. Hydrocarbon concentrations corrected for methane are determined by subtracting the methane value from the total hydrocarbon value.

Gravimetry

Gravimetry is the method of measuring air pollutants by direct weighing. Particulate matter is most commonly measured using this method. The procedure is as follows: particulates are removed from a known volume of ambient air by filtration or impaction; and the quantity of particulate matter collected is weighed. The total weight of material collected divided by the volume of air sampled is equal to the pollutant concentration on a weight basis (*i.e.*, $\mu g/m^3$). The most common application of the gravimetric technique is in the measurement of total suspended particulates (TSP) using the high volume (Hi-Vol) sampler.

REFERENCES

1. "Designation of Unacceptable Analytical Methods of Measurement for Criteria Pollutants," EPA Publication OAQPS No. 1.2-018 (May 1974).

2. "Methods of Air Sampling and Analysis," Intersociety Committee, American Public Health Association, 1015 Eighteenth Street, NW, Washington, D.C. (1972).
3. Elfers, L. A. "Field Operating Guide for Automatic Air Monitoring Equipment," IPA Publication No. PB-202-249 (July 1971).

CHAPTER X

AIR MONITORING INSTRUMENT CALIBRATION

INTRODUCTION

The objective of this chapter is to present an overview of the procedures required to calibrate continuous gas analyzers. This includes a discussion of the purpose of calibration procedures, reference methods of calibration and methods of generating known concentrations of pollutant gas mixtures. The importance of frequent calibrations as the best method of providing quality assurance for air monitoring data is the basic premise of this chapter.[1] Calibration determines the relationship between the instrument measurement and the true value of a specific pollutant. Since present-day air monitoring instruments are complex and subject to electronic drift, as well as time variation in such factors as sensor and reagent response, both an initial calibration of new instruments and periodic re-calibration are necessary to assure the accuracy and validity of the monitoring data obtained over a period of time. Accurate, reliable data become increasingly important as more significance is placed on small changes in the trend of pollutant levels and as the pollutant concentration approaches legally established air quality standards. Furthermore, air quality data gathered from different sources can be compared only if all the analyzers have been properly calibrated using accepted, reproducible techniques. According to Tokiwa and deVera,[2] calibration is performed in order to:

1. verify data
2. check instrument operation and detect defective components
3. establish the response of the analyzer's detection and signal presentation components to allow adjustment of the instrument operating parameters so the output can be made to conform to a predetermined curve or function
4. determine the frequency of calibration required by an instrument.

Calibration is a time-consuming, involved process which is most efficiently performed by trained and experienced personnel. To minimize expense and instrument down-time, the analyzer should be checked prior to calibration for evidence of incipient or existing malfunction, unless the purpose of the calibration is to determine the effect of such malfunction on previous data. A thorough air monitoring instrument calibration should include electronic checks and liquid and gas flow calibrations as well as the calibration of the instrument detector using several known concentrations of air pollutant. Not all instrument calibration procedures need to be this extensive, however. In order to draw distinctions between extensive calibration procedures, and rapid, less sophisticated calibration methods, two levels of instrument calibration procedures are discussed: primary calibration and secondary calibration.

METHODS OF CALIBRATION

Primary Calibration

A primary calibration involves the introduction of gas samples of known composition to an instrument operating side-by-side with the reference method for the purpose of producing a calibration curve or adjusting the instrument to match the reference method. This curve is derived from the instrumental response obtained by introducing several successive samples of different known concentrations. These standard gas mixtures are introduced in an increasing order of concentration to avoid contamination of the inlet lines and to minimize response times. The number of reference points necessary to define this relationship depends on the nature of the instrument output.

A primary air monitoring instrument calibration includes a thorough examination of the performance of the instrument electronics, a calibration of liquid and gas flow systems and a dynamic calibration of the instrument detector using reference methods. Figure 70 shows a flow diagram for the dynamic calibration procedure, performed with an artificially generated pollutant concentration. A major component of this dynamic instrument calibration is the simultaneous measurement of the gas stream by both the instrument and the reference method. This is accomplished using a dilution panel, which allows a concentrated stream of a specific pollutant gas to be diluted with clean air in varying amounts. Figure 71 shows a schematic diagram of a typical dilution panel. There are two inlet connections, one for a concentrated source of a desired pollutant gas and the other for diluent (*i.e.*, dilution) air. Before mixing with the pollutant gas, the diluent air is cleaned by columns

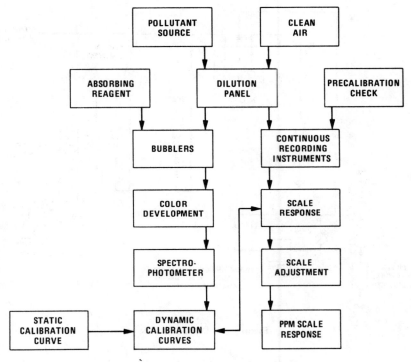

Figure 70. Dynamic instrument calibration.

containing activated carbon and a desiccant.   Each air flow is accurately measured by rotameters with micrometer adjustments before being combined in a mixing chamber.   The diluted mixture then passes simultaneously to the reference method and to the analyzer.   A wide range of concentrations of the calibration gas can be produced by varying the ratios of the two flows into the mixing chamber.

The reference methods represent the "primary standard" measurement to which the instrument being calibrated is compared.   Since reference methods (see *Federal Register* 36FR8186) are predominantly wet, chemical midget impingers are shown in Figure 71.   In the case of carbon monoxide, the midget impingers should be replaced with an NDIR CO analyzer when calibrating another carbon monoxide measuring instrument. Table 9 lists EPA reference methods for each of the criteria pollutants.

Secondary Calibration

A secondary air monitoring instrument calibration is less reliable than a primary calibration, but is much easier and quicker to perform.   Thus,

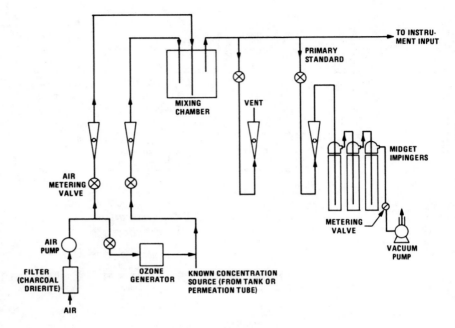

Figure 71. Dilution panel.

a secondary calibration can be performed much more frequently and conveniently than a primary calibration, while still providing important information regarding the instrument response characteristics.

A secondary calibration usually consists of a simple two-point check for instrument response: (a) check response of the instrument to "clean" air containing no air pollutants (zero gas) and (b) check response of the instrument to a single known concentration of pollutant gas in air (span gas). The span gas should be of high enough concentration to produce an instrument response equal to from 50% to 80% of full scale. This type of calibration determines the response of the instrument to (a) no air pollution in the sample ("instrument zero") and (b) the magnitude of the response per part per million of air pollutant present (instrument "span factor," measured as percentage of full-scale response per ppm of pollutant).

Secondary calibration procedures may or may not employ EPA reference methods as a primary standard. However, if the secondary calibration method used does not use EPA reference methods, then the calibration method should produce results equivalent to the EPA reference method. Thus, before a secondary calibration procedure can be used to check an

Table 9. Recommended Methods for Calibration

Pollutant	Federal Reference Method for Calibration[a]	Test Gas Generation Method[b]	Verification[b]
Photochemical oxidants (corrected for SO$_2$ and NO$_2$)	KI colorimetric	Ozone generator[c]	Gas-phase titration[d]
Carbon monoxide	Certified gas cylinder	Cylinder of CO in air or nitrogen	NBS certified[e]
Nitrogen dioxide	NaOH colorimetric	Permeation tube[f,g]	Gas-phase titration[d]
Nitric oxide	—	Cylinder of 100 ppm NO in prepurified nitrogen	Gas-phase titration[d]
Nonmethane hydrocarbons	Certified gas cylinder	Cylinder of methane in zero air	NBS certified[e]

aReference methods for calibration are the methods recommended in the *Federal Register* 36FR8186 for calibrating reference methods of measurement of criteria pollutants.

bPreferred methods of generating test atmospheres as published in the *Federal Register* 40FR7053 (February 18, 1975).

cAs described in *Federal Register* 36(228) (November 25, 1971).

dAs described in section 7.1 of *Federal Register* 38(110) (June 8, 1973).

eIf NBS (National Bureau of Standards) standards are not available, obtain two standards from independent sources which agree within 2%; or obtain one standard and submit it to an independent laboratory for analysis which must agree within 2% of the supplier's nominal analysis.

fAs described in Reference 3.

gAs described in Reference 4.

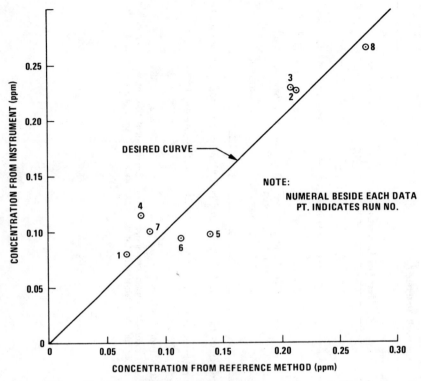

Figure 72. Concentration from instrument versus concentration from reference method.

instrument, it must undergo a primary calibration of its own. In this way, all instruments and all secondary calibration methods are linked back to the primary calibration which usually would use an EPA reference method.

CALIBRATION PROCEDURE

The procedure for performing both primary and secondary calibrations is outlined in Figure 73. Before the detector calibration can begin, the instrument's electronics are checked and the liquid and gas flow systems calibrated. Known pollutant gas concentrations, generated using either permeation tubes, gas generators or certified high pressure gas cylinders can then be diluted with pollutant-free air to the desired concentration for calibration of the instrument.

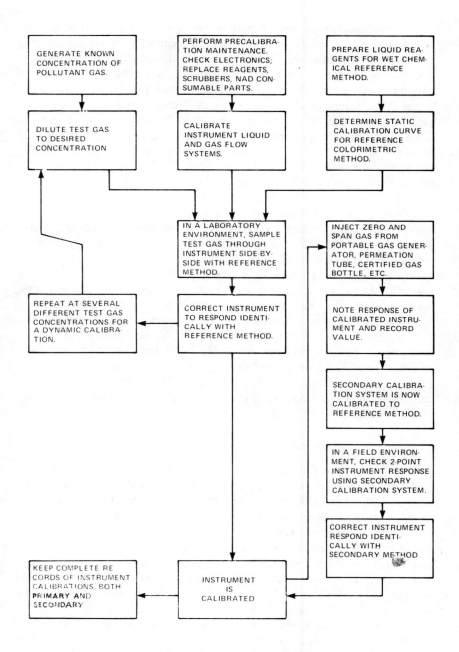

Figure 73. Outline of primary and secondary calibration procedures.

Both the instrument being calibrated and the reference method should sample the test gas simultaneously. The concentration of air pollutant measured by the reference method and the instrument being calibrated can then be compared. If the instrument does not indicate the same concentration as the reference method, the instrument output is adjusted to equal the results of the reference method determination. In order to perform a dynamic calibration, the procedure should be repeated several times at different air pollutant concentrations. After the instrument output is adjusted so that it is equivalent to the reference method at all pollutant concentrations, then the instrument is considered to be calibrated using primary calibration procedures.

Once the instrument has been calibrated, then the secondary calibration procedure can be calibrated also. This is achieved by injecting an air sample from the secondary calibration apparatus into the calibrated instrument. (Secondary calibration equipment may include portable permeation tubes, O_3 generators, certified gas bottles or a gas mixture in an air sampling bag.) The response of the instrument is recorded for both zero gas and span gas from the secondary calibration apparatus. These values are then used for further secondary calibrations. The secondary calibration procedure is then calibrated with the primary calibration procedure.

Complete records of instrument calibrations should be kept with the instrument. Both primary and secondary calibrations should be recorded, as well as any maintenance performed on the instrument or adjustments made to the instrument to correct response. These maintenance and calibration records become an important source of information regarding quality assurance.

Frequency of Calibration

The frequency with which an air monitoring instrument needs to be calibrated is primarily dependent on the operational stability of the instrument, which varies among different types of equipment and under different environmental conditions. Most present-day monitoring instruments are subject to drift and variation in internal parameters and cannot be expected to maintain accurate calibration over long periods of time. This is especially true when instruments are operated under field conditions where room temperature, humidity and the exposure to sunlight are not carefully controlled. Therefore, it is necessary to check and standardize operating parameters on a periodic basis.

For the best results, the frequency of calibration should be determined by observing the instrument under actual monitoring conditions. Frequent calibrations after the instrument is initially installed will allow for an estimation of the instrument's drift rate. The monitoring instrument should be recalibrated within the period of time required to drift a percentage of full scale of the instrument output. This may mean very frequent calibration (*i.e.,* hourly) for unstable instruments operating under uncontrolled field conditions. Calibration once every several days may be sufficient for stable equipment operating in environmentally controlled rooms.

In most cases, frequent routine calibrations are of the secondary type, employing a two-point calibration procedure (upscale span and zero setting). Since instrument zero, or baseline values, tend to drift or fluctuate more than the instrument span factor, the frequency of calibration will usually be based on the rate of "zero drift." Generally, when low ambient pollution levels are being measured, zero drift will cause more measurement error than "span drift;" therefore, frequent calibration of the instrument zero value is important.

Even when a good secondary calibration program is employed, primary calibration should be performed on a routine basis (*i.e.,* monthly or quarterly) on continuously operating equipment. Primary calibrations are also advised whenever new equipment is operated or when old equipment is repaired. Under these circumstances it is important that a dynamic calibration be performed in order to check the linearity of instrument response. Primary calibrations are also required periodically to recalibrate the secondary calibration apparatus.

ELEMENTS OF CALIBRATION

Precalibration Check

Before performing a primary instrument calibration, it is advisable to make sure that the instrument is in top working order. This is best accomplished by following the "instrument startup" instructions provided by the manufacturer in the instrument operation manual. Some of the obvious checks which should be made are outlined by Tokiwa and deVera:[2]

Supplies

1. fresh and adequate supply of reagent
2. reagent recovery system, if any (*i.e.,* charcoal column) recently renewed
3. adequate supply of recorder chart paper and supplies

Operation

1. function of instrument components (detector, pumps, recorder, etc.) appears normal

2. air and reagent flow rates recently validated and appear correct
3. instrument clean with no evidence of leaks in plumbing system
4. past data appear normal.

Liquid and Gas Flow Calibrations

Most automatic monitors have gaseous flow systems that must be maintained at a fixed rate for optimum operation. These systems must be maintained, calibrated and adjusted to the manufacturer's recommended specifications to provide accurate service. Two methods for gas flow calibration are in general laboratory use: (a) primary calibration using a wet-test meter, and (b) secondary calibration using a precalibrated mass flow meter. The use of the mass flow meter results in more rapid measurements, but it is subject to instrumental drift and should be checked periodically with a wet-test meter. All other flow metering devices, such as rotameters, should be checked against a wet-test meter. A more rapid approach to insuring the performance of mass flow meters in routine usage is to reserve one calibrated mass flow meter at the central laboratory as a reference and check the field meters against it before use. The test meter is placed upstream to the instrument's intake system in order to maintain atmospheric pressure through the device. Several data points (at least 5) are obtained by varying the control valve on the instrument. The volumetric flow data is plotted on linear graph paper against the rotameter setting to obtain a flow rate calibration curve. A typical rotameter calibration data sheet is presented in Figure 74. A typical flow rate calibration curve is presented in Figure 75.

Static Calibration

Static calibration is a performance test of the detection and signal presentation components accomplished by using an artificial stimulus—such as standard calibrating solutions, resistors, screens, optical filters or electrical signals—which has an effect equivalent to pollution concentrations. This procedure, which checks only the measuring systems of the analyzer and not the efficiency of absorption columns or accuracy of reagent or air flow rates, is used to establish the response of the detection and signal presentation components to permit adjustment of the instrument operating controls and parameters. Calibration with solutions is generally not feasible with instrument detection systems which do not circulate liquid reagents or whose design does not allow introduction of calibrating solutions.

INSTRUMENT _____

ROTAMETER SERIAL NO. _____

CALIBRATED WITH _____

LOCATION _____

TEMPERATURE _____ °C

ATMOSPHERIC PRESSURE _____ mmHg

RELATIVE HUMIDITY _____ %

CALIBRATED BY _____

TEST POINT	ROTAMETER READING	TOTAL FLOW, LITERS	TIME MIN.	FLOW RATE LITERS/MIN.
1				
2				
3				
4				
5				
6				
7				
8				
9				
10				

Figure 74. Rotameter calibration data sheet.

Reference Methods for Calibration

The reference method for three of the six criteria pollutants—NO_2, O_3 and SO_2—is the wet chemical colorimetric technique. The remainder of the gaseous pollutants (hydrocarbons and carbon monoxide) use certified concentrations of cylinder gas as a means of standardization.

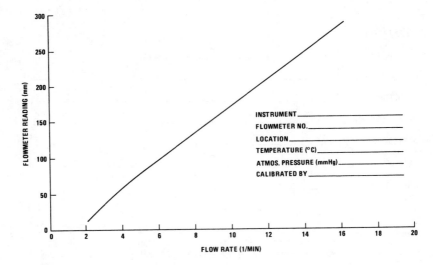

INSTRUMENT_____
FLOWMETER NO._____
LOCATION_____
TEMPERATURE (°C)_____
ATMOS. PRESSURE (mmHg)_____
CALIBRATED BY_____

Figure 75. Typical rotameter calibration curve.

Certified concentrations in steel cylinders can be used because of the relative stability of both CO and CH₄ gases in low concentrations for long periods of time. However, certified concentrations of these pollutants are not always reliable. Therefore, it is recommended that span gases certified by the National Bureau of Standards (NBS) be purchased to be used as an "audit" calibration gas. An "audit gas" is used to check the accuracy of another "working gas" used for frequent instrument calibrations. In this way, each calibration can be linked back to the NBS certification.

In the colorimetric method, the gas stream is bubbled through a liquid absorbing reagent. The color developed in the reagent follows Beer's law, and a spectrophotometer can be used to measure the degree of absorption due to this color development. Since the color development is a function of the amount of specific pollutant, the two must be related through the production of a static calibration curve.

A working standard solution, an absorbing reagent, and sometimes an indicator are combined to produce the static calibration curve. A working standard solution is a substance that has a precisely defined or calculated response when added to the absorbing reagent. The volume (ml) of the working standard solution can be related stoichiometrically to the concentration of the specific pollutant, in micrograms or microliters. Therefore, the static calibration curve can be obtained by varying the amounts of the working standard added to the absorbing reagent and by determining the spectrophotometer response. This linear curve provides a relationship between the amount, μg or μl, of the pollutant collected in the absorbing reagent and absorbance units.

In order to obtain the concentration of the gas absorbed, the color developed in the reagent is measured with a spectrophotometer. The measured absorbance, optical density, is used with the static calibration curve to obtain the amount of the specific pollutant. Then the actual concentration can be determined by dividing the amount of the pollutant, μg or μl, by the total volume of gas passed through the bubbler.

Generation of Known Gas Concentrations

Introduction

Basically, there are three commonly used methods of generating known air pollutant concentrations at levels suitable for air monitoring instrument calibration: permeation tubes, gas generators, and certified calibration standards contained in high pressure steel cylinders. EPA has recently published (40FR7053) preferred methods for generating test atmospheres and suggested methods of verifying the concentrations (see Table 9). For example, ozone should be produced using a gas generator, carbon monoxide and methane should be obtained in compressed gas cylinders, and NO_2 and SO_2 should be produced using permeation tube apparatus. Each of these three methods has been described by Elfers.[5]

Permeation Tubes

A permeation tube consists of a liquefied material contained under its own vapor pressure in a sealed section of permeable fluorinated ethylene propylent (FEP Teflon) tubing. The gradual permeation of the enclosed material through the FEP Teflon tubing permits the dispensing of microgram quantities of that material. Following a "conditioning" period of a few hours to several weeks, permeation proceeds at a highly constant rate until the enclosed material is nearly exhausted. Although the rate of permeation is highly dependent upon temperature, it is independent of normal changes in pressure and composition of the atmosphere.

To prepare specific permeation tubes for use in calibration of SO_2 and NO_2 monitors, pure liquid SO_2 or NO_2 is contained in a tube of FEP Teflon and sealed at the ends with Teflon plugs. Figure 76 is a diagram of a permeation tube.

The tubing is selected with diameter and length such that the permeation rate of the gas that it contains can be determined gravimetrically to

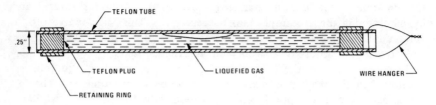

Figure 76. Permeation tube.

at least three significant figures over a reasonable period of time. Techniques used in making the permeation tubes and charging them with gas are described by O'Keefe and Ortman.[3]

Calibration of a liquid-filled tube consists of collecting weight data losses over a period of weeks. Between weighings, the tube must be maintained at a controlled temperature slightly above ambient, usually $25 \pm 0.1°C$, and a low humidity using silica gel or a comparable desiccant. Weight losses per unit of time are expressed as permeation rates.[4,6]

The following equation permits calculating parts per million (V/V) pollutant in a gas flowing over a tube as a function of air flow rate:

$$C = \frac{PR}{M} \times \frac{MV}{L}$$

where: C = ppm (V/V) pollutant transferred as a gas flowing over the tube
  PR = permeation rate at $25°C$ in $\mu g/min$
  M = molecular weight
  MV = molecular volume at $25°C$ (24.45)
  L = air flow in liters/min

Permeation Tube Dilution Apparatus

Figure 77 illustrates an apparatus employing permeation tubes for producing controlled low-level concentrations of standard gas mixtures. The system employs a gravimetrically calibrated permeation tube and purified air from which pollutants and moisture have been removed by passing through scrubbing columns filled with 8- to 12-mesh activated charcoal and a desiccant.

The diluent air is passed through a needle valve and a precision rotameter, calibrated within ± 1% by a wet test meter, to control and measure the flow. A purge flow of purified air (approximately 0.5 liters/min) is passed through the temperature conditioning coil and over the permeation tube, which is held in a Pyrex glass holder, submerged in

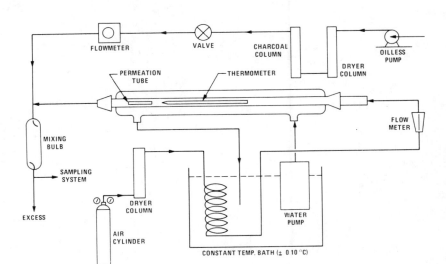

Figure 77. Permeation tube dilution apparatus.

a water bath, and controlled to 25 ± 0.1°C. After the diluent air and the purge gas stream are mixed in the mixing bulb, the resultant pollutant concentration is varied by adjusting the needle valve to control the air stream flow rate.

A portable calibration apparatus using permeation tubes has been developed for field application (see Figure 78). Since the apparatus was designed specifically for portability, it does not include a thermostatically controlled water bath. The general specification, construction details and use of this apparatus have been described by Rodes et al.[7] Commercial portable calibration systems are also available.

Standard Gases

Mixtures of stable gases, prepared to exact concentrations in pressurized cylinders, can be used as standard calibration gases following a confirmatory analysis by a referee laboratory. Commercially available mixtures of methane in air or nitrogen and mixtures of carbon monoxide in air or nitrogen have been found to be relatively stable over periods of several months, especially when contained in wax-lined or aluminum cylinders. Span calibration of carbon monoxide and hydrocarbon monitors can be achieved by direct introduction of these standard gases to the monitors. Normally, a manifold arrangement such as that shown in Figure 79 is employed to maintain the appropriate sampling conditions.

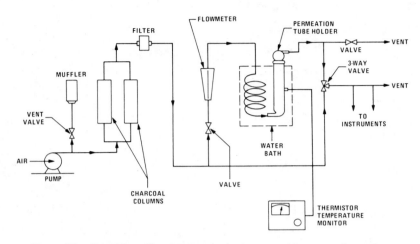

Figure 78. Portable calibration apparatus for use with permeation tubes.

Preparation of cylinder-stored standards of low concentrations of the more reactive gas mixtures such as sulfur dioxide or nitrogen dioxide in air cannot be achieved since these gases are unstable at the required concentrations. However, higher concentrations of SO_2 or NO_2, in the 0.5% range, have been found to be stable for several months providing inert diluent gas and clean storage cylinders are used. Mixtures of these gases can be diluted to lower calibration levels by means of a dilution panel device.[8],[9]

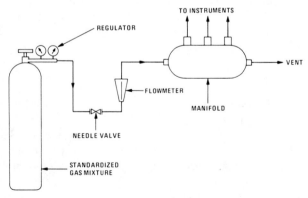

Figure 79. Distribution system for use with standard gases.

Gas Generation—Ozone

The ozone calibration apparatus consists of an 8-in. pencil-type mercury lamp that irradiates a 5/8-in. quartz tube through which clean air flows at 5-10 liters/min. By variable shielding of the lamp envelope, the generation of ozone may be varied from 0-1 ppm. The flow rate is controlled by a needle valve and measured by a rotameter (Figure 80).

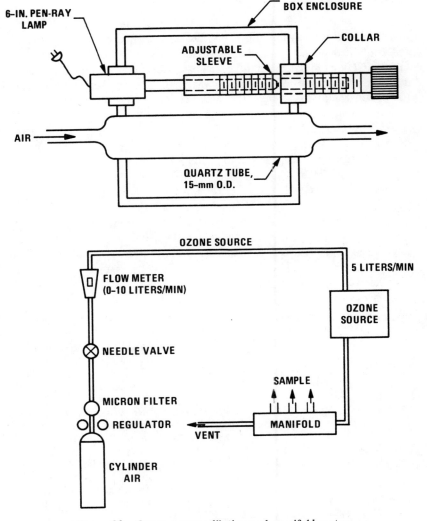

Figure 80. Ozone source, dilution and manifold system.

The ozone air passes to a manifold from which the monitor under test draws its sample. Standard impingers are used to sample the test atmosphere. A certified ozone generator is available from the National Bureau of Standards.

Wet chemical analysis of the sample collected simultaneously in the manual bubbler provides data for the preparation of the calibration curve.

Preparing Gas Mixtures in Sample Bags

Another method of dynamic calibration commonly used prior to the advent of the permeation tube employs standard gas mixtures prepared and contained in inert, impervious plastic bags. Polymeric plastic films such as Teflon, Mylar, Tedlar, Scotch-pak, Saran and Cellophane have been used for this application. Of these materials, Teflon, Mylar and Tedlar have been found to be the most desirable for overall applications.

Gas mixtures are prepared in bags by metering a large volume of diluent gas into the bag and then injecting a small amount of concentrated gas from a syringe through a serum stopper attached to the inlet fitting. Mixing is accomplished by kneading the bag. Prior conditioning of the bag for several hours with the same magnitude of concentration of pollutant is necessary to avoid losses through adsorption of the walls of the bag.

If the volumes of diluent and sample gases have been accurately measured, and if the mixtures are stable for the particular gas mixture bag combination selected, the standard can be assumed to contain its calculated concentration. The concentration should be verified by a referee wet chemical method.

REFERENCES

1. Noll, K. E., J. O. Walling and B. L. Arnold. "Calibration Procedures for Continuous Air Monitoring Instruments," Vol. I, Report No. 73-5, Submitted to Tennessee Department of Public Health, by Environmental Engineering Program, Dept. of Civil Engineering, University of Tennessee, Knoxville (September 1973).
2. Tokiwa, Y. and E. R. deVera. "Outline for Workshop on Static and Dynamic Calibration of Continuous Analyzers," Air and Industrial Hygiene Laboratory, State Department of Public Health, Berkeley, California.
3. O'Keefe, A. E. and G. C. Ortman. "Primary Standards for Trace Gas Analysis," *Anal. Chem.* 38(760) (1966).
4. Scaringelli, F. P., S. A. Frey and B. E. Saltzman. "Evaluation of Teflon Permeation Tubes for Use with Sulfur Dioxide," *Amer. Ind. Hygiene Assoc. J.* 28(260) (1967).
5. Elfers, L. A. "Field Operations Guide for Automatic Air Monitoring Equipment," EPA Publication No. PB 202-249 (July 1971).

6. Scaringelli, F. P., A. E. O'Keefe, E. Rosenberg and J. P. Bell. "Preparation of Known Concentrations of Gases and Vapors with Permeation Devices Calibrated Gravimetrically," *Anal. Chem.* 42(871) (1970).

7. Rodes, C. E., J. A. Bowen and F. J. Burmann. "A Portable Calibration Apparatus for Continuous Sulfur Dioxide Analyzers," 62nd Annual Meeting of the Air Pollution Control Association, St. Louis, Missouri (June 1963).

8. Chrisman, K. F. and K. E. Foster. "Calibration of Automatic Analyzers in a Continuous Air Monitoring Program," presented at the Annual Meeting of the Air Pollution Control Association, Detroit, Michigan (June 1963).

9. Nishikawa, K. "Portable Gas Dilution Apparatus for the Dynamic Calibration of Atmospheric Analyzers," presented at the Fifth Conference on Methods in Air Pollution Studies, Los Angeles, California (January 1963).

CHAPTER XI

AIR MONITORING HARDWARE

INTRODUCTION

The purpose of this chapter is to identify the type of equipment required to perform air quality monitoring. This includes not only air pollution analyzers but also plumbing, pumps, regulators recorders and other supplemental equipment. The first section discusses various types of air samplers, followed by sections on continuous air monitoring instruments, calibration equipment, data recorders, sample preconditioners, probes and tubing, flow control and measurement, and air movers. Tables are included in each section presenting typical price ranges and commercial suppliers of the equipment discussed.[1-5]

AIR SAMPLERS

Bubblers

The impinger, bubbler and wetted column are among the most widely used air sampling devices. With this type of sampling a chemical solution is used to remove the pollutant specie from the ambient air sample and fix or stabilize the pollutant for subsequent analysis in the laboratory. In some cases, chemically interfering substances may be removed prior to absorption of the pollutant in the sampling solution.[6]

Grab Sampling Vessels

Cold traps, plastic bags or glass syringes may be used to obtain ambient air samples for analysis. Cold traps concentrate volatile pollutants by using coolants such as dry ice or liquid nitrogen. Samples collected in plastic bags or syringes are analyzed directly in the laboratory. Bag

161

samples integrated over sampling periods of from five minutes to one hour are usually collected using battery-operated pumps with timing devices.[6]

Table 10. Commercial Sources of Air Sampling Equipment

Equipment Types	Typical Price Ranges	Commercial Suppliers
Bubblers & impingers	$10-$50 each	ACE, RAC, SGI, LGI, Pyrex
Bag samplers (single or multiple)	$300-$1000	
Sampling bags	$2-$20 each	JCS, EM, CII
High-volume samplers	$200-$500	BGI, GMU, ICN, PS, SC, UNICO
Tape samplers	$200-$1000	RAC, PS, UNICO, MS, LS

Particulate Samplers

Filtration is the most common method for collecting particulate air pollution samples. Particles collected on filter papers can be weighed to determine the total particulate concentration, or the particles can be analyzed microscopically or chemically. High-volume samplers and tape samplers are widely employed for particulate air pollution sampling, while electrostatic and thermal precipitators find some special applications in particulate sampling. Inertial impaction collectors, such as cascade impactors or Anderson samplers, are frequently used for particle size distribution analyses.

CONTINUOUS AIR MONITORING INSTRUMENTS

State-of-the-Art

At the present time, continuous analyzers are available for the measurement of ambient concentrations of carbon monoxide, carbon dioxide, nitric oxide, nitrogen dioxide, sulfur dioxide, total hydrocarbons, methane, oxidants, ozone and hydrogen sulfide. Although a large variety of automatic analyzers are now available commercially, only a few have been properly field tested to determine their limitations, reliability and durability and to provide information about possible interferences.

Since methods and instruments for measuring air pollutants must be carefully selected, evaluated and standardized, Hochheiser et al.[7] recommend that the following eleven performance characteristics be carefully evaluated when choosing continuous air pollution analyzers:

1. Specificity—does the method respond only to the pollutant of interest in the presence of those other substances likely to be encountered in samples obtained from ambient air or pollution sources?

2. Sensitivity and range—is the method sensitive enough over the pollutant concentration range of interest?

3. Stability—is the sample stable? Will it remain unaltered during the sampling interval and the interval between sampling and analyses?

4. Precision and accuracy—are results reproducible? Do they represent true pollutant concentration in the atmosphere or source effluent from which the sample was obtained?

5. Sample averaging time—does the method meet the above-stated requirements for sample-averaging time of interest?

6. Reliability and feasibility—are instrument maintenance costs, analytical time, and manpower requirements consistent with needs and resources?

7. Zero drift and calibration—is instrument drift over an unattended operation period of at least three days slight enough to ensure reliability of data? Are calibrating and other corrections automatic?

8. Response, lag, rise and fall time—can the instrument function rapidly enough to record accurate changes in pollutant concentration that occur over a short time in the sample stream being monitored?

9. Ambient temperature and humidity—does the instrumental method meet all of these requirements over temperature and humidity ranges normally encountered?

10. Maintenance requirements—can the instrument operate continuously over long periods with minimum down-time, maintenance time and maintenance cost? Is service for the instrument available in the area of use?

11. Data output—does the instrument produce data in a machine-readable format?

Commercial sources and price ranges for continuous air monitors for measuring oxidants, carbon monoxide, total hydrocarbons, methane and nitrogen oxides are listed in Table 11.

CALIBRATION EQUIPMENT

The calibration hardware required for performing dynamic instrument calibrations includes a dilution panel, bubbler trains, chemical reagents,

Table 11. Commercial Sources of Continuous Air Monitoring Equipment[1]

Principle of Operation / Manufacturer	Model	Measuring Range (ppm)	Cost (Kilo $)
Ambient CO Monitors			
Electrochemical Cell			
EnviroMetrics Inc.	NS-300, C-328 C-128	9.1-10,000	3-4
Electrochemical			
Energetics Science Inc.	2100-2800	0.5-2000	1
NDIR			
Beckman Instruments, Inc.	864	2.5 ppm-100%	2
Beckman Instruments, Inc.	865	0.25 ppm-100%	3-4
Bendix Corp./Process Instruments Division	8500-5A	0.5-30	4-5
Calibrated Instruments Inc.	SL/LC	[a]-50%	3
Horiba Instruments Inc.	APMA-10	0.5-1000	[a]
Mine Safety Appliances	202	0.8-ppm-10%	3-4
Fluorescence NDIR			
Andros Inc.	7000	0.2-200	7
Hg Substitution-UV Absorption			
Bacharach Instrument Co.	US 400 L	0.05-500	3
Ambient THC-CH_4-CO Monitors			
Gas Chromatography-Flame Ionization Detection			
Analytical Instrument Development, Inc.	511	[a]	4
Analytical Instrument Development, Inc.	514	[a]	4-5
Beckman Instruments, Inc.	6800	0.01-[a]	7-8
Bendix Corp./Process Instruments Division	8200	0.005-100	7
Byron Instruments, Inc.	233A	0.01-300	6-7
Hewlett-Packard Co.	5781A	[a]	5
Varian Aerograph	1440,2440	0.05-[a]	[a]
Ambient Oxidant Monitors			
Colorimetric			
Monitor Laboratories, Inc.	8100A	0.005-2.5	2
Technicon	Monitor IV	0.0015-0.2	7
Amperometric			
Beckman Instruments, Inc.	908	0.004-1.0	3-4
Freeman Laboratories, Inc.	A102	0.002-10,000	4-5

Table 11, Continued

Principle of Operation Manufacturer	Model	Measuring Range (ppm)	Cost (Kilo $)
Ambient Oxidant Monitors			
Amperometric			
Intertech Corp.	Picos	0.05-10	5-6
Mast Development Co.	724-2	0.003-1.0	1
Philips Electronic Instruments	Multi-component	0.005-3.0	a
Welsbach Ozone Systems Corp.	H-100-LC	0.0002-3	1
Chemiluminescence			
AeroChem Research Laboratory, Inc.	AA-3	0.002-1000	6-7
	AA-4, AA-5	0.005-10	7-9
Analytical Instrument Development, Inc.	560	0.001-1.0	2-3
Beckman Instruments, Inc.	950	0.001-5.0	3
Bendix Corp./Process Instruments Division	8002	0.001-1.0	4
Kimoto Electric Co. Ltd	803,804	0.001-1.0	2-3
McMillan Electronics Corp.	1100	0.001-10	2-3
McMillan Electronics Corp.	1110	0.01-10,000	2-3
Meloy	OA 300	0.001-10	3-4
Monitor Labs	8410	0.001-5	2-3
REM Scientific, Inc.	612 B	0.002-2.0	4-5
Thermo Electron Corp.	12A	0.0001-2500	7
Ultraviolet Absorption			
Dasibi Corp.	1003-AH	0.003-20	3-4
Dasibi Corp.	8000	0.01-30	2
2nd Derivative UV Absorption Spectroscopy			
Spectrometrics/Lear Siegler Inc.	IIId^2	0.025-10	1
Ambient NO$_X$ Monitors			
Colorimetric			
CEA Instruments	PM 102	0.0025-4	2
CEA Instruments	PM 112	0.0025-4	2
CEA Instruments	PM 113	0.0025-4	2
CEA Instruments	PM 124/125	0.0025-4	3
Enraf-Nonius	NO	0.005-0.25	a
Enraf-Nonius	NO$_2$	0.005-0.25	a
Freeman Laboratories, Inc.	A100-A1001	0.002-10,000	3-6
Kimoto Electric Co. Ltd.	212	0.005-1.0	3-4
Monitor Laboratories, Inc.	8100A	0.005-2.5	2
Precision Scientific Co.	Aeron	0.02-2.0	2
Scientific Industries, Inc.	80	0.01-2.0	3
Technicon Industrial Systems	Monitor IV	0.0015-0.15	7
Xonics, Inc.	411 series	0.005-2.0	3-5

Table 11, Continued

Principle of Operation Manufacturer	Model	Measuring Range (ppm)	Cost (Kilo $)
Ambient NO$_x$ Monitors			
Amperometric			
Beckman Instruments, Inc.	909	0.004-1.0	3-4
Beckman Instruments, Inc.	910	0.004-1.0	3-4
Intertech Corp.	Picos	[a]-10	5-6
Mast Development Co.	724-11	0.1-2500	1
Philips Electronic Instruments	Multi-component	0.015-100	[a]
Electrochemical Cell			
Dynasciences Corp.	NX-110	0.1-100	2-4
Dynasciences Corp.	NR-210	0.02-100	2-4
Dynasciences Corp.	NS-410	0.02-10	6-7
EnviroMetrics, Inc.	NS-300, N-322 N-376, N-376M	[a]-10,000	2-5
Theta Sensors, Inc.	LS-400	0.03-3.0	2
Chemiluminescence			
AeroChem Research Laboratories, Inc.	AA-1, AA-2, AA-3	0.002-1000	6-7
AeroChem Research Laboratories, Inc.	AA-4, AA-5, AA-6	0.05-10	7-9
AeroChem Research Laboratories, Inc.	RS-2	0.001-10,000	7-8
Beckman Instruments Inc.	952	0.005-25	5-6
Bendix Corp./Process Instruments Division	8101B	0.005-5	6
Intertech Corp.	—	[a]	[a]
LECO Corp.	CL-30	0.001-10,000	[a]
McMillan Electronics Corp.	1200	0.004-2.0	4-5
Meloy Laboratories, Inc.	520	0.005-5.0	6
Monitor Labs, Inc.	8440	0.002-5000	[a]
REM Scientific, Inc.	642	0.005-10	6
Thermo Electron Corp.	12A	0.0001-2500	7
Thermo Electron Corp.	14B	0.0005-10	5-6
Ambient Hydrocarbon Monitors			
Flame Ionization Detection			
Antek Instruments, Inc.	840	0.01 ppm-5%	2
Beckman Instruments, Inc.	400	0.02 ppm-5%	2
Bendix Corp./Process Instruments Division	8400	0.01-1000	2
Bendix Corp./Process Instruments Division	8201	0.005-50	5
Delphi Industries	C4A1	0.1-10,000	[a]
Gow-Mac Instrument Co.	23-500	0.07 ppm-100%	2

Table 11, Continued

Principle of Operation Manufacturer	Model	Measuring Range (ppm)	Cost (Kilo $)
Ambient Hydrocarbon Monitors			
Flame Ionization Detection			
Meloy Laboratories, Inc.	SH 202-2	0.1-5000	5
Mine Safety Appliances	11-2	a-20	5-6
Mine Safety Appliances	Total hydrocarbon	a	2
Power Designs Pacific,Inc.	1562	0.13-1000	4
Process Analyzers, Inc.	30-100	0.06 ppm-10%	2
Scott Aviation	11-6500	a	a
Teledyne Analytical Instruments	400	0.02-10,000	3-8
Thermo Electron Corp.	30	a	a

aInformation not available.

flow regulators, permeation tubes, gas generators and certified pollutant gas mixtures in compressed gas cylinders. This equipment is adequately described in Chapter X and will not be further discussed here. However, Table 12 does list commercial sources of calibration equipment as well as the approximate costs of some of the important components.

Table 12. Commercial Sources of Air Monitoring
Calibration Equipment

Equipment Type	Typical Price Range	Commercial Suppliers
Dilution panel	$800-$2000	TAI, ML, CII, UNICO
Bubbler trains	$40-$100	RAC, ACE, SGI, UNICO
Chemical reagents	Varies widely	WSI, FSC, AHT
Pressure regulators	$40-$90	WGP, LSG, APC, PGP
Permeation tubes	$30-$110	NBS, AID, MET, PC
Ozone generators	$600-$1800	REM, ML, BEN
Bottled gas mixtures	$25-$240	MGP, LSG, APC, PGP

DATA RECORDERS

Most air pollution monitoring instruments employ detectors or sensors which produce a variable DC voltage electrical signal that is proportional to the concentration of the pollutant being measured; the signal can be measured using either an analog or a digital volt meter. In most cases, however, air monitoring instruments require continuous electronic data recording devices such as analog strip chart, digital magnetic and paper tape recorders to record the output from the air pollution monitoring instruments (see Figure 81). Output from the air monitoring sensors can

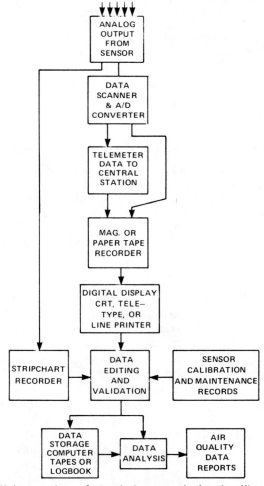

Figure 81. Unit operations of a typical aerometric data handling program.

input into an electronic data scanner and analog to digital converter. In digital form, the signal can be recorded on magnetic or paper tape recorders. When data telemetry systems are used, the signal from the converter may be transmitted long distances to a central data receiving station where the data is recorded. At the central station, digital data displays such as a CRT (cathode ray tube), a teletypewriter or a line printer may be employed in order to obtain a visual record of the data as it is received.

Raw data from strip chart or digital tape recorders should be edited and validated with the help of instrument calibration and maintenance records. Upon completion of data editing and validation the data can be stored for later use in a computer data bank, on digital tape, on key punch cards, or as "hard-copy" in a data logbook. Air quality data storage, analysis and report preparation can be performed either manually or with the assistance of computers. Some sophisticated data acquisition systems are completely computerized, except for the data editing and validation step where human judgment is needed for best results.

Strip Chart Recorders

Strip chart recorders employ a moving paper chart and a marking pen to produce a continuous visible recorder of the variable DC signal output from an air monitoring instrument. The deflection of the marking pen on the strip chart can be related to the ambient air pollution concentration via a calibration curve, thus providing a visual record of the pollutant concentration fluctuation with time. Peak pollutant concentrations as well as time-averaged concentrations can be read manually from a strip chart. Strip chart recorders are available with different chart sizes, chart speeds, number of data channels, voltage input ranges, pen types, power requirements and costs. Chart widths typically range from about 5 to 30 cm (2 to 12 in.) with chart speeds of from 1 cm/hr to 10 cm/sec: the width and speed determine the resolution obtainable from the data record, with wide charts and fast speeds providing the highest resolution.

Recorder marking pens may use colored inks, or special chart papers which are sensitive to either the pressure or the temperature of the pen. Heat-sensitive and pressure-sensitive chart paper is more expensive than conventional paper, but it is often more reliable and produces a more uniform record than inking systems. This is especially true of thermal recording systems. Inking systems usually cause a considerable amount of data loss due to problems with ink supplies clogging, inking too heavily, and pens tearing the chart paper. Most inking strip chart recorders require frequent maintenance for proper operation.

The cost of strip chart recorders varies widely depending on the size and special features offered by the manufacturer. Small, single-channel, milliamp current recorders sell for as little as $100, while a good two-channel, potentiometric recorder with variable chart speeds and multiple voltage input ranges may cost $2500 (see Table 13 for commercial sources).

Table 13. Commercial Sources of Data Acquisition Equipment

Equipment Type	Typical Price Range	Commercial Suppliers
Strip chart recorders	$100-$2000 per channel	EA, LN, SEC, HPC
Punch tape recorders	$1000-$4000	DSC, DEC
Electronic data loggers	$2000-$10,000	EA, ML, WE, HPC
Magnetic tape recorders	$1000-$7000	DSC, DEC, HPC, HON
Minicomputers	$500-$2500 per 1000K core	DSC, MSI, DEC, HON

Data Scanners and A/C Converters

The first step in digital data recording is to take the analog output from an air monitoring instrument and convert the analog signal to its digital equivalent. This is usually accomplished using an electronic device which accepts the analog output from several different air pollution sensors, scans each of these sensors in a sequential order, multiplexers between channels and converts the analog signal to digital form. The data from each scan of the device is recorded sequentially on a digital magnetic tape or paper tape recorder along with the date and time that the sensors were scanned. The primary components of this type of digital data logger are illustrated in Figure 82.

Several important design specifications of digital data loggers are scan rate, scan interval, number of available input channels and the binary code used by the analog to digital (A/D) converter. The scan rate which is usually limited by the recording medium employed, is the speed at which the data scanner accepts information from the different sensors, and is usually adjustable from speeds of less than one to as many as several hundred channels per second. The scan interval, which is adjustable in most devices, is the time period between sensor scans; typical scan intervals range from one scan per second to one per hour. The scan rate and the scan interval determine the amount of data collected and the rate at which recording tape will be used.

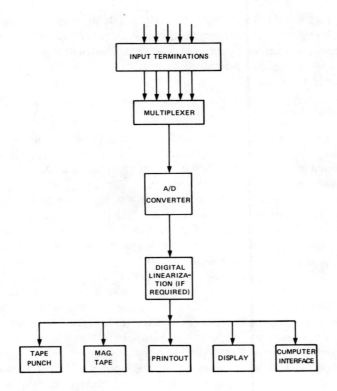

Figure 82. Principal components of a digital data acquisition system.

The design of A/D (analog to digital) converters and multiplexers varies considerably among different manufacturers. An especially important characteristic, which should be noted when purchasing A/D converters, is the binary code output of the digitizer. Typical commercial data loggers use binary codes consisting of from four- to eight-bit coding languages. Using four- and six-bit binary codes allows the data loggers to be built at lower costs because fewer electronic components are required. The simpler binary languages also require less recording space on digital data recorders. A significant disadvantage of these systems results when the information is to be input to large business type computers or read into standard teletypewriters or line printers. These more sophisticated systems commonly employ the U.S.A. Standard Code for Information Interchange, USASCII-70, and eight-bit binary code. Thus, data loggers using four- and six-bit binary codes usually require an electronic data translator to convert to an eight-bit code which is "computer compatible." The additional cost of the electronic translator may equal or exceed the initial cost of the data logger. Costs of data loggers applicable for air quality data acquisition may range from $2,000 to $10,000 (for commercial sources see Table 13).

Digital Data Recorders

Most digital data recording is performed utilizing either magnetic tape recorders or punch-paper tape recorders depending on the application and the preference of the user. Digital output from an electronic data logger can be recorded directly on either device.

Magnetic tape recorders store binary data in the form of variations in the polarity of magnetized iron oxide particles impregnated in plastic, paper or metal tape. Magnetic tape recorders vary according to the width of the magnetic tape used, taping speeds, number of data channels or "tracks" and the type of tape drive used; they can be of either the "write only" or "write and play-back" variety. Recorders are available that use reel-to-reel, cartridge and casette tape drives. Prices range from $1200 for a simple system to $7000 for a high-quality magnetic tape recorder. Some commercially available data loggers even have self-contained tape recorders eliminating the need for a separate unit (for commercial sources see Table 13).

Punched paper tape recorders utilize holes punched in a continuous strip of plastic or paper tape as an information storage mechanism. Each row of the paper tape consists of an index hole, used to locate the row along the length of tape, and five-, six-, seven- or eight-hole positions (depending on the binary code used) representing the data. In a punch tape recorder, data transferred from the data logger is temporarily stored in what is called the punch buffer. When the index hole on the tape is sensed by the recorder, the contents of the punch buffer are gated to the punch magnets and the binary data bits are punched in the tape. While the tape is moved to the next row position, incoming information is transferred from the data logger to the punch buffer and the process is repeated.[8]

Punched tape systems are generally less expensive than magnetic tape systems. However, they also provide less flexibility and less data space per reel of tape, which means that paper tapes must be replaced more frequently than a similar-sized reel of magnetic tape, and the paper tape sometimes breaks or tears in the machine resulting in data losses. In systems where this is a problem, mylar tapes may be used in place of paper. Punched tape recorders typically cost from $1000 to $4000 (for commercial sources see Table 13).

Computer-Controlled Data Systems

A network of continuous monitoring stations can be operated under the control of an on-line computer located at a central facility. When a remote station receives a command to transmit, a scanning device interrogates each sensor and assembles the information into a message which is then transmitted to the receiver at the central station where it is immediately input to the computer. The on-line control unit at the central station may be either a small computer totally dedicated to functioning as a control module or a time-shared general purpose computer.

When instantaneous sensor outputs are transmitted to the central station, the on-line computer accumulates the values for each sensor over an interval of time (usually one hour or less) and computes a time-averaged pollutant concentration. The average concentrations are then printed to obtain a hard copy, which provides a continuing record of air quality. When the basic time-averaged concentrations are calculated, they are also written on computer-compatible magnetic tape, which is processed by the computer to update the master data files. Typical mini-computer with 8,000 bytes of usable core cost from $6,000 to $10,000 (for commercial sources see Table 13).

SAMPLE PRECONDITIONING

Some air monitoring equipment requires preconditioning of the air sample before introduction to the air pollution analyzer. This preconditioning is usually designed to remove substances from the sample that might interfere with the analysis while not affecting the pollutant under investigation. Sample preconditioning may consist of steps such as particle removal, humidification of the sample, drying of the sample, heating or cooling the sample, or selectively scrubbing an interfering chemical specie. Particle removal is achieved using in-line filters ahead of the air pollution analyzers. Humidification is achieved by bubbling the air sample through water. Drying air requires either a cooler-condenser mechanism, or drying columns containing solid desiccants such as silica gel or calcium sulfate. Heating the air sample requires that heating tape be wrapped around the outside of the sample lines; cooling the sample is best achieved by immersing the sample tubing in a water or ice bath. Selective scrubbers require special solid reagents packed in a column through which the air sample must be drawn (for commercial sources and price estimates of these various types of air sample preconditioning hardware see Table 14).

Table 14. Commercial Sources of Air Sample Preconditioning Equipment

Equipment Type	Typical Price Ranges	Commercial Sources
Bubblers	$40-$100	RAC, UNICO, ACE, SGI
Filters and holders	Varies widely	MIL, GEL, RA, MSA, RAC
Electrical heating tape	$1-$10/ft	CPI, ACE, SGI, CUR
Drying tubes	$5-$25 each	CPI, ACE, MSS, CUR
Desiccants	$2-$5/lb	CPI, FSC, MSS

SAMPLING PROBES AND TUBING

The sampling probe is a tube within which the ambient air sample is transported from outside the monitoring station to the air pollution analyzers. The critical elements of sampling probe design are proper tubing size and construction materials. Sufficient air must be drawn through the sample probe for all the air monitoring instrument, while the retention time of the air sample within the sampling manifold and tubing should be as short as possible. This helps to insure that the air sample reaches the analyzer without undergoing chemical alterations (Table 15).

Table 15. Commercial Sources of Air Sampling Tubing

Equipment Types	Typical Price Range		Commercial Sources
Glass tubing	1/4" OD	$4-$6/100 ft	
	1/2" OD	$10-$15/100 ft	
Tygon tubing	1/4" OD	$15-$20/100 ft	EHS, MSS, WSI,
	1/2" OD	$40-$50/100 ft	FSC, ACE, CPI,
Polyethylene tubing	1/4" OD	$3-$7/100 ft	(Sources same
	1/2" OD	$10-$15/100 ft	for all tubing)
Teflon tubing	1/4" OD	$80-$125/100 ft	
	1/2" OD	$125-$150/100 ft	

FLOW CONTROL AND MEASUREMENT

In order to obtain quantitative analytical results from most air pollution measuring methods it is of upmost importance that the volume or volumetric flow rate of air being sampled is known precisely and accurately. This is because most analytical methods are selective for the air pollutant being measured sensing only the volume or mass of pollutant present. The concentration of pollutant in the air sample is determined by dividing the volume of pollutant measured by the volume of ambient air sampled, yielding results in parts per million concentration by volume (ppm). For this reason, most air pollution analyzers employ flow regulating and flow rate measuring hardware.

Air Flow Rate Regulation

Depending on the application, sample air flow rates can be controlled by using either a pump with a constant flow rate output or by using a pump that produces a constant air pressure or vacuum output with a restrictor in the sample lines providing a controlled resistance to fluid flow.

Constant Volume Pumps

Pumps that produce a constant volumetric air flow rate are of the positive displacement type. Reciprocating piston pumps and some types of rotary lobe pumps (i.e., a Roots blower) can be used for this purpose. In a positive displacement pump, the same volume of air is drawn into the pump on each rotation or reciprocation of the pumping mechanism. Therefore, the output flow rate of the pump can be controlled by regulating the speed at which the pump operates. This is usually accomplished by controlling the electrical voltage applied to the pump motor using a variac or rheostat. The speed of the motor can be measured using a tachometer. This type of pump requires periodic calibration, after which a constant flow rate output can be obtained by electrically controlling the voltage applied to the pump motor. Most applications of constant flow volume output pumps are for flow rates in the several CFM (cubic feet per minute) range.

Flow Control by Resistance

The more widely used method of controlling air flow rate is a vacuum pump and controlled resistances in the air sample lines (see Figure 83). The vacuum pump draws air through the sample lines into the sample

Figure 83.   Flow rate control mechanism.

collection device and through a flow measuring device before entering the pump. In most air moving devices the flow rate decreases as the resistance it must overcome increases. Therefore, the air flow rate through the system can be regulated by using dampers, adjustable valves (usually needle valves), or calibrated orifices to add resistance to the sample lines. The flow resistors are usually installed immediately before the vacuum pump, so that the sample collection device and flow measuring device operate at lower vacuum pressures than the maximum produced at the pump. This is especially critical for flow measuring devices that are designed to operate at near atmospheric pressures. The amount of resistance produced depends on the size of the valve or orifice used. Adjustable resistances can be obtained by partially opening or closing a needle valve. When an orifice is used the flow rate is controlled by changing the size of the orifice. Hypodermic needles are widely used as critical flow orifices in air sampling equipment. Table 16 lists the size of hypodermic needle required to control air flow rates between 0.06 and 14.0 liters per minute.

Flow Control by Diversion

In some varieties of air sampling equipment, air flow regulation is improved by using flow diversion methods as well as resistance. Flow diversion simply means that the air drawn through the vacuum pump is not all pulled through the sample collection device. To achieve this, a "bleed valve" is introduced in the sampling train as shown in Figure 84. By adjusting the amount of "bleed-in air," the air flow through the sample collection device can be increased or decreased. Bleed valves are frequently used with diaphragm pumps in order to produce more uniform flow characteristics through the sampling train.

Air Flow Measurement

Methods used to measure air sample flow rate fall into three categories: (a) gas velocity measurements, (b) volumetric flow rate measurements,

Table 16. Performance of Hypodermic Needles as Critical Orifices[a2]

Gauge	(cm)	Flow Rates (liters/min) at Needle Length (cm)		
		8.9	5.1	2.5
13		14		
15	0.14	10		9.9
17			6.9	
18			4.4	5.2
19	0.069		2.8	2.8
20			2.1	2.3
21			1.4	1.7
22			0.90	1.0
23	0.032		0.49	0.63
24				0.49
25			0.21	0.30
27	0.019			0.16
30	0.015			0.06

[a]760 torr, 20°C upstream

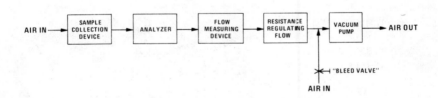

Figure 84. Flow rate control by diversion.

and (c) total volume measurements. Gas velocity measurements are most applicable for the measurement of large volumes of air such as found in flues, ducts or in the atmosphere (wind speed measurements). Pitot tubes, spinning cup anemometers and thermal anemometers all find applications in gas velocity measurements. Because of the large volumes usually associated with gas velocity measurements, these methods are infrequently used in ambient air monitoring equipment. Volumetric flow rate meters and total volume measuring devices are more commonly employed.

Total Volume Measurement

The method universally considered as the "primary standard" for air flow measurement is to measure the total volume of air drawn through the air sampler, divided by the time required to pump the measured air volume. The total gas volume can be measured using either a spirometer, a water displacement bottle, a soap bubble meter, a wet test meter or a dry gas meter. The time required to pump the air volume is usually measured with a stopwatch. The type of volume measuring device used depends on the volume to be measured and the accuracy desired.

Spirometer

The spirometer is one of the simplest and most dependable methods of measuring gas volume, and is usually used as a primary standard for calibrating other types of air flow meters. The spirometer consists of a cylinder of known volume, closed at one end, with the open end submerged in a circular tank of water (see Figure 85). The cylinder is

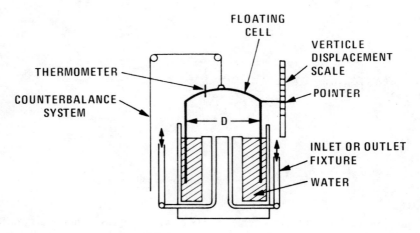

Figure 85. Spirometer.

counterbalanced with weights so that the air pressure inside the cylinder can be closely controlled. As the air being measured enters the spirometer, water is displaced and the inverted cylinder rises. A pointer on a graduated scale indicates the volume of water that has been displaced which is equal to the volume of gas measured. Spirometers are available with cylinder volumes ranging from 0.2 ft^3 to 20 ft^3 (for commercial sources see Table 17).

Table 17. Commercial Sources of Flow Control and Measurement Equipment

Equipment Types	Typical Price Ranges	Commercial Sources
Needle valves	Varies widely	CUR, CPI, ACE
Hypodermic needles (orifices)	Varies widely	GCI, HC, UUC, RAC
Wet test meter	$300-$650	AMC, PS, CUR
Dry gas meter	$150-$300	AMC, IA, CUR
Spirometer	$125-$300	AMC, WEC, AHT
Soap bubble meter	$50-$150	SKC, TH
Volumetric glassware	Varies widely	WSI, FSC, AHT
Rotameter	$40-$200	FPC, BI, RGI, CPI
Venturi meter	$50-$200	ICN
Orifice meter	$50-$200	EI

Water Displacement Bottles

When a spirometer is not available, a very similar procedure can be employed using a water displacement bottle. In this method, air is bubbled into a large bottle containing water (see Figure 86). As the air enters, an equal volume of water is displaced and removed from the bottle. The bottle can be graduated or the volume of water removed can be measured using graduated volumetric laboratory glassware. This method is less accurate than the spirometer because the air pressure inside the bottle varies during the measurement. Displacement bottles are used only for low air flow rates.

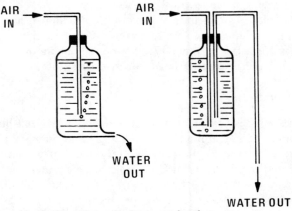

Figure 86. Water displacement bottles.

Soap Bubble Meter

Another device for measuring gas volumes is the soap bubble meter. It consists of a cylindrical graduated glass cylinder (*i.e.,* a laboratory buret) open at both ends (see Figure 87). The air sample can be either

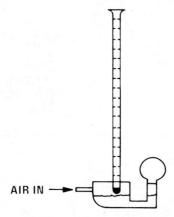

AIR IN

Figure 87. Soap bubble meter.

forced or drawn through the cylinder. The direction of flow is usually from the bottom to the top. As the air volume being measured is drawn through the cylinder, a soap bubble is introduced at the bottom of the cylinder. The bubble will follow the column of air as it moves up the glass cylinder and can be timed as it passes graduated markings. The air volume sampled is equal to the volume of the cylinder displaced by the bubble. This method is quite accurate, though applicable only for small gas volumes and gas flow rates. Commercial sources of soap bubble meters are listed in Table 17; however, a laboratory buret and a bottle of soap bubbles represents an effective substitute.

Wet Test Meter

The wet test meter consists of a series of inverted buckets of traps mounted radially around a shaft and partially immersed in water (see Figure 88). The location of the entry and exit gas ports is such that the entering gas fills a bucket, displacing the water and causing the shaft to rotate due to the lifting action of the bucket full of air. The entrapped air is released at the upper portion of the rotation and the bucket again fills with water. In turning, the drum rotates the index

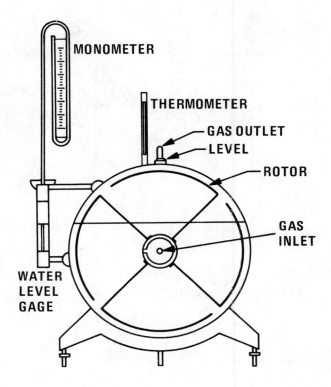

Figure 88. Wet test gas meter.

pointers which register the volume of gas passed through the meter. The wet test meter is usually considered as a secondary calibration standard for gas volume measurement. It is available in a range of sizes capable of measuring from 10-1200 cubic feet per hour (for commercial sources of wet test meters see Table 17).

Dry Gas Meter

The dry gas meter employs two or more movable partitions, or diaphragms, attached to the case by a flexible material so that each partition may have a reciprocating motion (see Figure 89). The gas flowing into one bellows chamber inflates it, actuating a set of slide valves which shunt the incoming flow to another bellows. The inflation of the successive chambers also actuates a set of dials which register the volume of gas passed through the meter. Dry gas meters are usually used for measuring large gas volumes of relatively large air flow rates and are

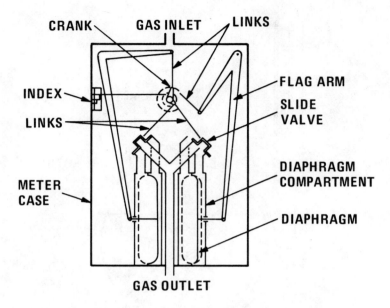

Figure 89. Dry test gas meter.

available in sizes ranging from 20 to 1800 cubic feet per hour (for commercial sources of dry gas meters see Table 17).

Flow Rate Measurement

For most applications, direct measurement of the air flow rate (liters or cc per minute) through an air sampler are the most convenient. Basically there are two principles of operation for gas flow rate measuring devices: (a) differential head meters, which are used primarily on high velocity gas streams, and (b) differential area meters, which are used over a wide range of gas flow conditions.

Differential Head Meters

Differential head meters are those in which the stream of fluid passing through a constriction in the flow meter creates a significant static pressure difference which can be measured and correlated to the volumetric flow rate through the meter. The shape of the constriction is usually either an orifice or a venturi shape, hence, these meters are usually referred to as orifice meters or venturi meters (see Figure 90). Two pressure taps, one upstream and one downstream of the constriction,

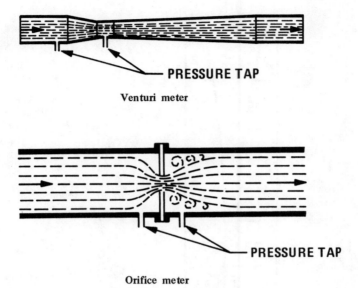

PRESSURE TAP

Venturi meter

PRESSURE TAP

Orifice meter

Figure 90. Differential head meters (venturi and orifice).

serve as a means of measuring the static pressure head differential, or "pressure drop," across the device. The pressure drop is usually measured using a manometer or bellows-type pressure gauge (for commercial sources see Table 17).

Differential Area Meters

The differential area meter differs from differential head meters in that the pressure drop across an area meter remains constant while the cross-sectional area of the constriction changes with the volumetric gas flow rate. By far, the most common type of differential area meter is the rotameter. A rotameter consists of a vertical graduated glass tube, slightly tapered in bore, with the diameter decreasing downward (see Figure 91). Inside this tube is a specially shaped float, of diameter slightly greater than the minimum bore of the conical tube. The fluid flow to be measured passes upward through the conical tube until the float reaches a position in the tube where its weight is balanced by the force due to the fluid flowing past it. The position of the float in the tube depends on the gas flow rate.

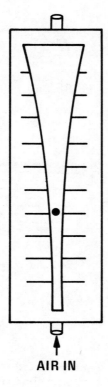

Figure 91.   Rotameter flowmeter.

AIR IN

Rotameters must be calibrated using a spirometer or wet test meter. A graduated scale is usually marked off on the glass tube of the rotameter. After calibration, this scale can be used to read directly the volumetric gas flow rate through the rotameter for different positions of the float. Rotameters are available in sizes ranging from 1 cc/min up to very large flow rates (for commercial sources see Table 17).

AIR MOVERS

Air movers are required to induce the flow of sample air through the sampling manifold, tubing, sample collector, flow measuring device, and the air pollution analyzer. In some systems only one air mover is required, while in others, separate air movers are needed for pulling air through the sampling manifold, into the sample collector, and into the pollution analyzer. Various types of air movers are employed for different applications, depending on the flow rate requirements, vacuum or pressure requirements, portability requirements, and the potential for sample contamination. Air movers range in moving capacity from a few cubic centimeters of air up to several cubic meters per minute; in operational complexity they range from a squeeze bulb to a multi-stage pump.

An air mover can be used to force an air sample into a sample collector using air pressure, or it can be used to induce an air sample into an analyzer, etc., using a vacuum. In most cases it is preferable to locate the air sampling device as the first unit in a sampling train and draw the air sample into the device using a vacuum pump located on the downstream side. This procedure eliminates possible contamination of the air sample by the pump, and allows for flexibility in the type of pump used. In some circumstances, however, the pump will have to be located upstream of the air sampling device. In this case, the type of pump used must be very clean and constructed of materials that will not contaminate or otherwise alter the chemical composition of air sample (*i.e.*, through adsorption or absorption of the pollutant gas). Hendrickson[4] recommends that the pump parts that contact the air flow should be fabricated of either polytetrafluorethylene or polychlorotrifluoroethylene, though other inert materials may also be suitable.

Motor-Driven Pumps

Basically there are four different types of electric motor-driven pumps commonly used in air pollution sampling: reciprocating piston pumps, reciprocating diaphragm pumps, rotary lobe or vane pumps, and centrifugal blowers. Each of these pumps has special applications depending on flow rate, vacuum and pressure requirements. The advantages and disadvantages of each type are compared in Table 18.

Reciprocating Piston Pumps

The principle of operation of a piston pump is like that of the internal combustion engine. Air is drawn into a cylinder chamber on the suction stroke of a piston and then is pushed out on the discharge stroke (see Figure 92). On the suction stroke, the suction valve is open, allowing air to flow in; on the discharge stroke, the suction valve closes, and the discharge valve opens, allowing air to flow out. Piston pumps vary in complexity of operation from a manually operated piston to a mechanically operated one with many working parts.

Reciprocating Diaphragm Pumps

The operation of a diaphragm pump is very similar in principle to a piston pump. The piston (plunger) in a diaphragm pump does not move

Table 18. Pump Comparison[3]

Pump Type	Advantages	Disadvantages
Piston pump (reciprocating)	1. Can operate at high suction pressure 2. Can be metered	1. Small capacity 2. Seal required between piston and piston chamber 3. Working parts such as check valves and piston rings may cause difficulties 4. Pulsating flow
Diaphragm pump (reciprocating)	1. Wide range of capacities 2. No seal required 3. Small capacity, battery operated units available	1. Limited materials of construction 2. Operation at limited suction pressures 3. Pulsating flow
Gear pump Lobe pump Vane pump (rotary)	1. Can be metered 2. Self-priming 3. Simple construction 4. Low maintenance costs 5. Pulseless flow 6. Wide range of capacities 7. Can operate at high suction pressures 8. Self-lubrication may be available	1. Close clearance—bad for air with high solids 2. Limited materials of construction 3. Slippage between movable parts 4. Pool self-lubrication at low vacuum
Centrifugal pump	1. Large range of capacities 2. No close clearance	1. No small capacities 2. Turbulence 3. Operational noise 4. Relatively low suction pressures

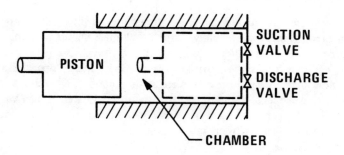

Figure 92. Piston pump.

in a tightly fit chamber as the piston pump; rather, it is attached to the center of a circular diaphragm, the outer edge of which is bolted to a flange on the pump casing.   The diaphragm may be made of metal or rubber—the most important characteristic being its flexibility and resistance to reaction with the air being moved.   The up-and-down motion of the plunger is permitted by diaphragm flexibility without the rubbing of one part on another (see Figure 93).   On upward movement of the plunger, air flows into the pump through a suction valve.   Downward movement of the plunger closes the suction valve, and the air is forced through a discharge valve, perhaps located in the plunger itself.

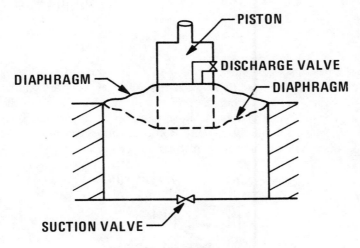

Figure 93.   Diaphragm pump.

Rotary Lobe Pumps

A lobe pump consists of two or more lobes enclosed in a closely fitted casing, arranged so that when the lobes mesh on one side, fluid (air) fills the resulting space between the lobes.   The air filling the space is carried around the pump to the opposite side and is displaced as the lobes again mesh.   Lobe pumps are produced with two, three, four or more lobes (see Figure 94).   Another common rotary lobe pump is the cycloid pump which is a specially shaped two-lobe variation.

Rotary Vane Pumps

A vane pump consists of one rotor in a casing, machined eccentrically in relation to the drive shaft.   The rotor contains either a series of

Figure 94.   Lobe pumps.

movable vanes, blades or buckets which follow the bore of the casing, thereby displacing fluid (air) with each revolution of the drive shaft. As the rotor turns, air flows from the intake into the space between the vane and the casing. The trapped air is transported around and forced out the discharge side of the pump. Rotors are produced using both swinging vanes and sliding vane configurations (see Figure 95).

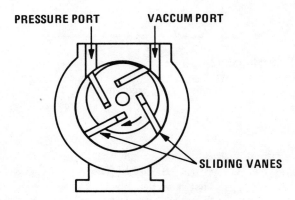

Figure 95.   Sliding vane pump.

Centrifugal Pumps

Centrifugal pumps (or blowers) employ centrifugal force to move air. The simplest form of this type of pump consists of an impeller rotating in a volute casing. The rotation of the impeller creates a decreased pressure at the impeller "eye" thus causing air to be drawn into the pump.

Air drawn into the center of the impeller is "picked up" by the vanes, accelerated to a high velocity by rotation of the impeller, and then discharged by centrifugal force into the casing and out the discharge. As air is forced away from the impeller eye more air is drawn in due to the vacuum created. Centrifugal pumps used in air sampling can be divided into three categories: radial flow (see Figure 96), axial flow and mixed flow.

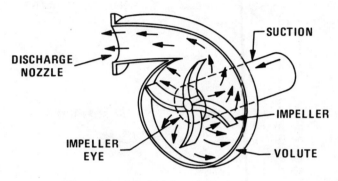

Figure 96. Centrifugal pump operation.

Performance Characteristics of Motor-Driven Pumps

Several operating characteristics of motor-driven pumps are of special interest: the flow capacity of the pump, suction pressure, the materials of construction, whether the air flow is continuous or pulsating, the ability of the pump to maintain a constant flow rate, and the electric power consumption of the pump motor or transducer. Some of these characteristics, such as capacity, materials of construction, ability to maintain a constant flow rate, materials of construction, and power consumption vary for different sizes and makes of similar types of pumps, but these characteristics are usually described in the manufacturer's literature. Other characteristics of pumps, such as suction pressure versus capacity and whether the air flow is continuous or pulsating is dependent on the pump type. For example, a pulsating air flow is a characteristic of reciprocating pumps, while smooth continuous air flow can usually be expected from rotary pumps.

The two major classifications of pump type are positive displacement pumps and centrifugal pumps. Included in the positive displacement classification are reciprocating piston and diaphragm pumps, and rotary lobe and vane pumps. The performance characteristics for positive displacement pumps are quite different from centrifugal pumps. In general, positive displacement pumps can produce much higher suction pressures at low flow rates. Positive displacement pumps are often characterized by a linear relationship between suction pressure and pump capacity while centrifugal pumps exhibit a nonlinear relationship. Also, the efficiency of positive displacement pumps is highest near the maximum rated capacity of the pump, while centrifugal pumps are most efficient at flow rates well below the maximum rated capacity (see Figures 97 and 98). Table 18 summarizes important performance characteristics of popular motor driven air movers.

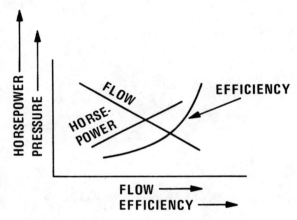

Figure 97. Characteristic curve for a rotary pump.

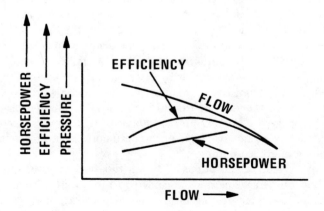

Figure 98. Characteristic curve for a centrifugal pump.

SHELTERS

When using air sampling equipment in the field it is imperative that the equipment be suitably housed in some type of shelter. The purpose of the shelter is multifold. Most importantly it provides protection from the weather which might otherwise damage the instrument. Second, many air monitoring instruments require constant temperature environments for stable operation, and this is best achieved using an air sampling shelter with thermostatically controlled heating and air conditioning. Third, it is also important to shield most air monitoring equipment from direct sunlight which can upset the instrument operation. Fourth, when large shelters are used, they may provide needed laboratory space for instrument calibrations, valuable work space for instrument maintenance and repair, and desk space needed for data reduction work. Finally, shelters can also guard the equipment from accidental damage or damage by vandalism.

When choosing a shelter for air pollution monitoring equipment there are several important characteristics that should be considered: size and construction, portability, durability, weatherproofing, plumbing and wiring and cost. The type of shelter will depend on the type of monitoring equipment to be used, the number of instruments to be operated in the shelter, and the mobility needed to accomplish air monitoring objectives.

The size and construction of an air monitoring shelter is usually the primary consideration. There are five basic categories of shelter sizes commonly used for air monitoring:

1. a "doghouse"-size shelter (10 sq ft)
2. an "out-house"- or mini-van-size shelter (40 sq ft)
3. a small trailer, step van, or portable building (120 sq ft)
4. a large trailer, motor home, or portable building (250 sq ft)
5. a permanent indoor laboratory (500 sq ft)

Each of these shelter types has specific air monitoring applications with inherent advantages and disadvantages. For example, the "doghouse"-size shelter is used to house a single air sampler (such as a multiple bag sampler), a gas bubbler or a single air monitoring instrument. This type of shelter would not normally have environmental controls (*i.e.,* heating and air conditioning) and it might or might not be wired for electricity. The shelter is portable and inexpensive (less than $200).

The "out-house"- or mini-van-size shelter provides more space and much more flexibility than the "doghouse" size. It can house two or three air monitoring instruments and recorders, and the inside of the shelter can be environmentally controlled using a small heater and air conditioner. Electric power is essential, but wet facilities are uncommon. This size shelter is quite portable, especially if a mini-van is used, and it costs from as low as $500 for an "outhouse"-size portable building to around $4000 for a mini-van.

Small trailers, step vans and comparibly sized portable buildings can provide room for several air monitoring instruments, recorders, and supplemental equipment. They are usually environmentally controlled inside to insure proper air monitoring instrument operation. Frequently electricity, water and telephone lines are connected to the station. Portability is high, especially with step vans. The costs of shelters of this size range from 1500 for a portable building to $6000 for a step van.

Large trailers, motor homes or large portable buildings can be used to provide space for a complete set of air monitoring instruments and data recorders and much supplemental equipment can be operated simultaneously within this type of shelter. Additional space is available for instrument calibration apparatus, gas bottle storage, sink, desk and toilet facilities. The environment inside the shelter is controlled using heaters and an air conditioner. These larger shelters are usually connected to electricity, water, sewer and telephone service. Portability is high for motor homes and trailers, but rather poor for large portable buildings. Costs of these shelters range from approximately $3000 for a large portable building to as high as $20,000 for a motor home.

Permanent buildings serve best as sites for permanent monitoring networks employing data telemetry systems. Air monitoring stations housed in permanent buildings have the same characteristics as large trailer or motor home installations. Permanent buildings are frequently used as the primary or central station of an air monitoring network consisting of several remote stations and one central station. Besides conducting air pollution measurements, the central station is used as a data acquisition center, and an instrument calibration and repair laboratory for the entire monitoring network.

KEY TO EQUIPMENT SUPPLIERS

ACE   ACE Glass Inc.
P. O. Box 688
Vineland, New Jersey 08360

AID   Analytical Instrument Development, Inc.
250 S. Franklin Street
West Chester, Pennsylvania 19380

AHT   Arthur H. Thomas, Co.
Vine Street at Third
P. O. Box 779
Philadelphia, Pennsylvania 19105

AMC   American Meter Company
13500 Philmont Avenue
Philadelphia, Pennsylvania 19116

APC   Air Products & Chemicals, Inc.
      Specialty Gas Division
      733 W. Broad Street
      Emmaus, Pennsylvania 18049

BEN   Bendix Process Instruments,
        Div.
      Drawer 477
      Ronceverte, West Virginia 24970

BGI   BGI Incorporated
      58 Guinan Street
      Waltham, Massachusetts 02154

BI   Brooks Instrument Div.
      Emerson Electric Co.
      407 W. Vice Street
      Hatfield, Pennsylvania 19440

CII   Calibrated Instruments, Inc.
      17 West 60th Street
      New York, New York 10023

CPI   Cole Parmer Instrument Co.
      7425 N. Oak Park Avenue
      Chicago, Illinois 60648

CUR   Curtin Scientific Co.
      P. O. Box 1546
      Houston, Texas 77001

DEC   Digital Equipment Corp.
      146 Main Street
      Maynard, Massachusetts 01754

DSC   Doric Scientific S. A.
      3883 Ruffin Road
      San Diego, California 92123

EA   Esterline Angus
      Box 24000
      Indianapolis, Indiana 46224

EHS   E. H. Sargent & Co.
      10558 Metropolitan Avenue
      Kensington, Maryland 20795

EI   Ellison Instrument Div.
      Dieterich Standard Corp.
      P. O. Box 96
      New Buffalo, Michigan 49117

EM   Environmental Measurements,
        Inc.
      215 Leidesdorff Street
      San Francisco, California 94111

FPC   Fisher & Porter Co.
      300 Warminster Road
      Warminster, Pennsylvania 18974

FSC   Fischer Scientific Co.
      Central Offices
      711 Forbes Avenue
      Pittsburg, Pennsylvania 15219

GCI   Glenco Scientific, Inc.
      3121 White Oak Drive
      Houston, Texas 77007

GEL   Gelman Instrument Co.
      600 South Wagner Road
      Ann Arbor, Michigan 48106

GMW   General Metal Works
      8368 Bridgetown Road
      Cleves, Ohio 45002

HC   Hamilton Co.
      P. O. Box 307
      Whittier, California 90608

HON   Honeywell Inc.
      1100 Virginia Drive
      Fort Washington, Pennsylvania
        19034

HPC   Hewlett Packard Co.
      P. O. Box 112
      Clayton, California 94517

IA   Instrumentation Assoc.
      17 West 60th Street
      New York, New York 10023

ICN   ICN Instrument Div.
      630-20th Street
      Oakland, California 64612

JCS   James Cox & Sons, Inc.
      P. O. Box 674
      Colfax, California
        95713

LGI Lab Glass, Inc.
1172 Northwest Boulevard
Vineland, New Jersey 08360

LN Leeds and Northrup
North Wales, Pennsylvania
19454

LS Leigh Systems, Inc.
220 Boss Road
Syracuse, New York   13211

LSG Linde Specialty Gases
223 Highway #18
East Brunswick, New Jersey
08816

MET Metronics Associates, Inc.
3201 Porter Drive
Stanford Industrial Park
Palo Alto, California 94301

MGP Matheson Gas Products
P. O. Box 85
East Rutherford, New Jersey
07073

MIL Millipore Corp.
Ashby Road
Bedford, Massachusetts 01730

ML Monitor Labs, Inc.
10451 Roselle Street
San Diego, California 92121

MS Monitoring Systems, Inc.
Core Laboratories
Box 10185
Dallas, Texas   75207

MSA Mine Safety Appliances Co.
201 N. Braddock Avenue
Pittsburg, Pennsylvania 15208

MSI Metrodata Systems, Inc.
617 Rock Creek Road
P. O. Box 1307
Norman, Oklahoma 73069

MSS Matheson Scientific
12101 Centron Place
Cincinnati, Ohio 45246

NBS National Bureau of Standards
Institute of Materials Research
Gaithersburg, Maryland 20234

PC Polyscience Corp.
909 Pitner Avenue
Evanston, Illinois 60202

PGP Precision Gas Products
P. O. Box 358
Linden, New Jersey 07036

PS Precision Scientific
3737 West Cortland Street
Chicago, Illinois 60647

Pyrex Pyrex Laboratory Glassware
Corning, New York

RA H. Reeve Angel & Co., Inc.
9 Bridewell Place
Clifton, New Jersey

RAC Research Appliance Co.
Rt. 8
Allison Park, Pennsylvania
15101

REM REM Scientific, Inc.
2000 Colorado Avenue
Santa Monica, California
90404

RGI Roger Gilmont Instrument,
Inc.
161 Great Neck Road
Great Neck, New York 11021

SC The Staplex Company
Air Sampler Division
777 Fifth Avenue
Brooklyn, New York 11215

SEC	Simpson Electric Co. 853 Dundee Elgin, Illinois 60120	UNICO	UNICO Environmental Instruments A. Gelman Company P. O. Box 590 Fall River, Massachusetts 02722
SGI	Scientific Glass & Instruments, Inc. P. O. Box 18306 Houston, Texas 77023	UUC	Unimetrics Universal Corp. 1853 Raymond Avenue Anaheim, California 92801
SKC	Schutte & Koerting Co. Instrument Division Cornwall Heights Bucks County, Pennsylvania 19020	WE	Westinghouse Electric Corp. Environmental Systems Center Meter Division Raleigh, North Carolina 27603
TAI	Tracor Analytical Instruments 6500 Tracor Lane Austin, Texas 78721	WEC	Warren E. Collins, Inc. 220 Wood Road Braintree, Massachusetts 02184
TH	Teledyne–Hastings Co. 333 W. Mission Drive San Gabriel, California 91776	WSI	Will Scientific, Inc. Box 20155, Station N Atlanta, Georgia 30325

REFERENCES

1. *Instrumentation for Environmental Monitoring–Air*, Lawrence Berkeley Laboratory, University of California, Berkeley (December 1973).
2. Mueller, P. K. "Detection and Analysis of Atmospheric Pollutants," in *Combustion Generated Air Pollution*, E. Starkman, Ed. (New York: Plenum Press, 1971).
3. "Analysis of Atmospheric Inorganics," Training Course Manual in Air Pollution, U.S. Dept. HEW, Public Health Service (July 1965).
4. Hendrickson, E. R. "Air Sampling and Quantity Measurement," in *Air Pollution, Vol. II*, A. C. Stern, Ed. (New York: Academic Press, 1968).
5. Elfers, L. A. "Field Operations Guide for Automatic Air Monitoring Equipment," EPA Publication No. PB-202-249 (July 1971).
6. Stern, A. C., H. C. Wohlers, R. W. Boubel and W. P. Lowry. *Fundamentals of Air Pollution* (New York: Academic Press, 1973).
7. Hochheiser, S., F. J. Burmann, and G. B. Morgan. "Atmospheric Surveillance–The Current State of Air Monitoring Technology," *Environ. Sci. Technol.* 5(8) (August 1971).
8. Considine, D. M. *Encyclopedia of Instrumentation and Control.* (New York: McGraw-Hill Book Co., 1971).

CHAPTER XII

PLANNING METEOROLOGICAL SURVEYS

INTRODUCTION

The effect of meteorology on ambient concentration of air pollution is both important and complex. The transport and dispersion of primary pollutant and photochemical formation of secondary pollutants are dependent on meteorology. In order to interpret variations in the temporal and spatial distribution of air pollution concentrations, a basic understanding of the role of the atmosphere in transporting and dispersing air pollutants is important. This requires a knowledge of basic meteorology as well as some more specialized air pollution meteorology.[1-5]

This chapter will identify the important meteorological parameters affecting the transport and dispersion of air pollution and describe methods of collecting and evaluating meteorological data. The uses of meteorological data in planning air quality monitoring studies are discussed, including the types of historical meteorological data that are available. Meteorological field studies are described including step-by-step procedures for conducting such studies. Guidelines on instrument selection, deployment, and exposure are presented.

THE USES OF METEOROLOGICAL DATA
IN ASSESSING AIR QUALITY

Introduction

Only after reviewing the types of meteorological conditions that occur throughout a study area can one hope to determine where, when and how often high ambient air pollutant concentrations are likely to occur. In many cases, an evaluation of available historical data taken at a U.S. Weather Bureau Station will be sufficient to allow any adequate analysis

197

of meteorological conditions, while in other cases the project will require meteorological monitoring.

Historical meteorological data, primarily measurements of wind speed and direction, are useful in analyzing the typical and worst case conditions affecting air pollution transport and dispersion within and outside the study area. Historical wind speed and direction data can be used to determine where air pollutants are likely to be transported, how these conditions change with time, and the frequency of occurrence of adverse dispersion conditions (i.e., low wind speeds and very stable atmospheric conditions) within the study area. This data can then be used to estimate when and where background air pollution concentrations and ambient pollution levels due to the proposed project will be high.

In most cases, meteorological data will be used to perform one of two types of analysis: (a) a mesoscale meteorological analysis, which is an evaluation of the area-wise air movements that affect background air pollution levels within the study area, and (b) a microscale meteorological analysis, which is an evaluation of local air movements that affect the dispersion of air pollutants near a project.

Mesoscale Meteorological Analysis

A mesoscale meteorological data analysis will include an evaluation of:

1. Prevailing surface wind patterns throughout the study area for different seasons of the year
2. When, how often and for what duration low wind speeds and stable atmospheric conditions occur
3. Meteorological conditions that transport air pollutants into the study area, resulting in high background air pollution concentrations
4. The effects of large-scale terrain features on the area-wide movements of air
5. Possible sites where additional meteorological data may be needed or where mesoscale air quality measurements should be made.

In order to properly evaluate each of these items, area-wide sources of air pollution and important terrain features should be considered. This can be accomplished by using a map to show the study area plus the surrounding topographic and land use features. Raised relief maps are especially useful in identifying significant mesoscale terrain features and identification of the following conditions:

1. Channeling effects of winds in canyons and valleys
2. Areas of drainage winds
3. General wind flow around hills and mountains for different inversion bases

4. Areas where a surface inversion may exist due to cool air drainage
5. Areas of limited dispersion due to valley walls

Sensitive receptors and existing and future major point, line and area sources of air pollution emissions should be included on the map. Since high local background air pollution concentrations usually occur downwind of major pollution sources, the map can be used to determine the direction from which the wind would transport air pollution into the study area. Adverse conditions for transporting pollutants out of the study area can also be identified. Historical meteorological data can then be evaluated to determine the frequency of occurrence of winds from each direction, the wind speeds and what seasons of the year and times of day adverse conditions usually occur. An air pollution transport analysis of this type should begin with the largest and closest air pollution source and continue until all major sources are evaluated. For a comprehensive analysis, sophisticated mathematical models may be used.[6,7]

A basic six-step methodology is recommended for performing a mesoscale analysis on historical meteorological data. This methodology, outlined in Figure 99, consists of the following steps:

1. Collect all available information including historical meteorological data, topographic data, demographic data, present and future land use data, present and future transportation plans, and available air pollution source emission inventory data.

2. Display as much of this information as possible on a topographic map; especially major transportation routes, major sources of air pollution, major land use characteristics, major terrain features, and the location of the proposed project and nearby receptors.

3. Prepare wind roses from historical meteorological data and plot on the map. (Plot at the location of meteorological station when possible.)

4. Determine probable wind flow streamlines for the most probable and "worst case" wind directions. Plot streamlines for different wind conditions on transparent map overlays.

5. Determine air trajectories which transport air pollutants into and out of the study area. (Air trajectories are most applicable for determining the potential for high concentrations of photochemical pollutants.)

6. Once adverse meteorological conditions for the transport of air pollutants either into or out of the study area have been identified, then a frequency analysis should be performed on the historical meteorological data to determine when (season and time of day), how often and for what durations adverse conditions occur. The frequency analysis should include an analysis of the frequency of occurrence of various wind speeds, wind directions and atmospheric stabilities, for different seasons of the year and times of the day.

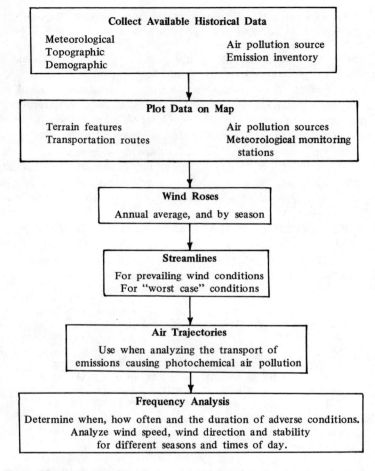

Figure 99.   Analyzing historical meteorological data for a mesoscale study.

Microscale Meteorological Analysis

The objective of the microscale evaluation is to study how the project influences the microscale movement of air and to determine how air pollutants generated by the project will be transported within and out of the study area.

A microscale meteorological analysis is probably best performed using a map including right-of-way boundaries, adjacent land use characteristics, nearby sensitive receptors, and microscale terrain features (hills, trees, buildings, etc.).   A microscale meteorological analysis can then be performed to attempt to identify potential problem areas for the transport and dispersion of air pollutants generated by the proposed project. A microscale meteorological data analysis will include an evaluation of:

1. The effects of local terrain features which may disrupt the flow of air within the study area
2. The expected flow patterns that pollutants will follow when being transported away from the project for different wind directions
3. Sensitive receptors located in a downwind direction of the project
4. When, how often and for what duration low wind speeds and stable atmospheric conditions occur
5. Possible sites where meteorological data may be needed or where microscale air quality measurements should be made

The methodology for performing a microscale meteorological analysis, which is basically the same as described for a mesoscale analysis, is outlined in Figure 100.

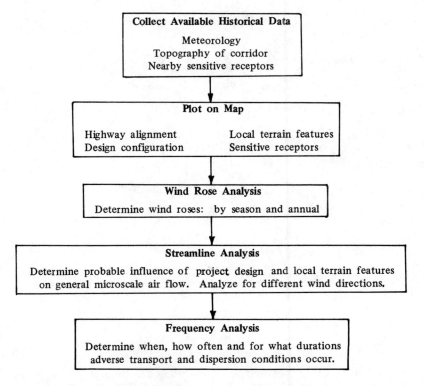

Figure 100. Analyzing historical meteorological data for a microscale study.

METHOD OF ANALYZING HISTORICAL METEOROLOGICAL DATA

Methods of analyzing historical meteorological data include wind roses, streamlines, air trajectories and frequency analysis. Wind roses are commonly used to display historical data from a single measuring station, illustrating the frequency of occurrence of winds from each of 16 wind directions. Sometimes, the frequency of occurrence of various wind speed ranges is also illustrated (see Figure 101). Streamlines, which are especially

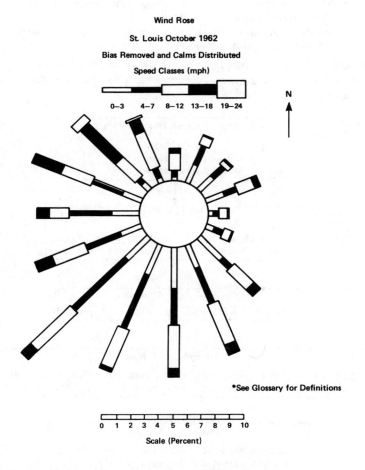

Figure 101. Example of wind rose presentation of wind speed and direction frequency data. (See Glossary for definitions.)

applicable when terrain features influence the general flow of air, are frequently used to illustrate both microscale and mesoscale air flow patterns under various prevailing meteorological conditions. Streamlines are especially useful for illustrating the mesoscale effects of channeling or different types of drainage and valley winds (see Figures 102 and 103). Air trajectories, which are used to evaluate the area-wide movement of a single "air parcel" as it is transported under changing conditions of

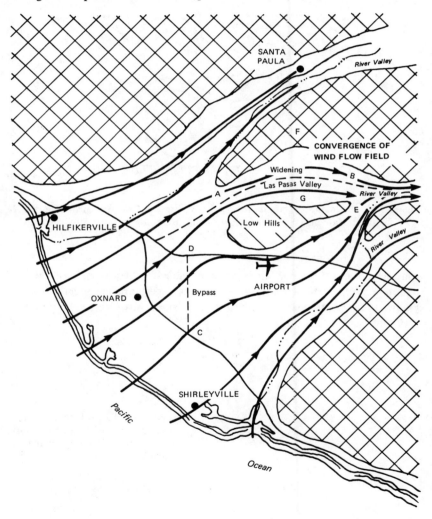

Figure 102. Example of a mesoscale streamline analysis. (From Beaton,[1])

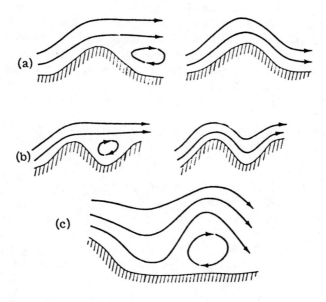

Figure 103. Examples of various flow patterns (streamlines) over microscale terrain features: (a) small hill, (b) hollow or dip, (c) rotor in the lee of a hill. (From Slade.[5])

wind speed and direction, are most applicable for "back-tracking" air pollutants from a final location and time to their original location (see Figure 104).   A frequency analysis is performed in order to determine

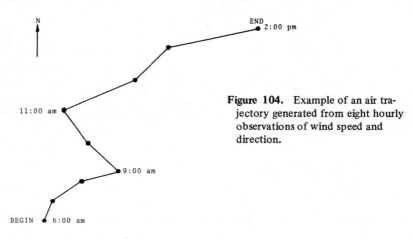

Figure 104.   Example of an air trajectory generated from eight hourly observations of wind speed and direction.

when and how often meteorological conditions occur that are adverse to the transport and dispersion of air pollutants (*i.e.,* an analysis of wind speed, wind direction, atmospheric stability). Using tabular representation of the results of a frequency analysis (see Table 19), the frequency of occurrence of adverse meteorological conditions can be determined for each hour of the day and season of the year.

AVAILABLE SOURCES OF METEOROLOGICAL DATA

Meteorological measurements of wind speed, wind direction, temperature, humidity, barometric pressure, ceiling height, visibility, precipitation, and some other parameters are recorded routinely each hour at over 300 U.S. weather stations. The raw data are permanently stored and are made available on request. This data can be obtained directly from the weather station taking the measurements or through the National Weather Records Center, Asheville, North Carolina. Raw data are available in log form, punch cards, magnetic tape, and, in some cases, as analog traces. Sometimes, only climatological summaries are needed. Several standard climatological summaries are available on request from the National Weather Records Center. A publication entitled *A Selective Guide to Published Climatic Data Sources Prepared by the U.S. Weather Bureau* and sold by the Superintendent of Documents, U.S. Government Printing Office, is an invaluable source of data summary information.

METEOROLOGICAL FIELD SURVEYS

Introduction

Besides the analysis of historical meteorological data, a project may require meteorological measurements within the study area. Field measurements may be desirable for two different purposes: meteorological data verification and air quality model validation. While the two types of studies may have many similarities, they usually differ appreciably in their objectives, types of equipment used, exposure criteria for instruments, duration of study, and methods of data evaluation. Meteorological data verification studies almost always require mesoscale meteorological measurements while air quality model validation studies may require both microscale and mesoscale meteorological measurements.

Meteorological Data Verification Studies

Meteorological measurements taken for the purpose of data verification are used to compare the observed conditions of meteorology within the

Table 19. Five-Year Meteorological Frequency Table (by season, wind direction, wind speed, stability and time of day)

Station	Airport																
Season	January, February, March																
Wind Direction	Northwest																
Wind Speed	< 2 m/s					2-3.5 m/s						4-6 m/s				> 6 m/s		Total
Stability	A	B	D	E	F	A	B	C	D	E	F	B	C	D	E	C	D	
Time 100			0.01	0.04	0.06					0.03	0.07				0.01			0.21
200			0.01	0.04	0.06					0.03	0.07				0.03			0.22
300			0.02	0.06	0.07					0.05	0.08				0.04			0.30
400			0.02	0.05	0.10					0.07	0.05							0.33
500			0.03	0.04	0.12				0.01	0.08	0.04			0.03	0.08			0.43
600			0.01	0.05	0.15				0.01	0.12	0.02			0.02	0.07			0.45
700			0.02	0.05	0.08				0.02	0.10	0.08			0.03	0.04		0.04	0.46
800			0.04	0.04					0.05	0.04	0.04			0.03	0.02		0.15	0.41
900			0.08					0.02	0.06				0.03	0.05			0.20	0.41
1,000			0.08					0.04	0.08			0.02	0.03	0.06			0.18	0.47
1,100		0.04	0.06				0.06	0.04	0.05			0.01	0.03	0.03		0.15	0.11	0.59
1,200		0.05	0.01				0.06	0.06	0.02					0.02		0.08	0.05	0.39
1,300	0.21	0.08				0.10	0.08	0.04	0.03			0.02	0.04	0.02		0.20		0.82
1,400	0.25	0.08				0.10	0.17	0.08	0.03			0.02	0.05	0.03		0.22		1.03
1,500	0.30	0.10				0.08	0.12	0.12	0.02			0.01	0.03	0.02		0.10	0.05	0.98
1,600	0.15	0.09				0.12	0.08	0.05	0.01			0.02	0.06	0.02		0.18	0.06	0.91
1,700		0.06	0.07				0.06	0.04	0.04			0.01	0.03	0.02		0.05	0.16	0.54
1,800			0.05					0.05	0.05				0.03	0.06		0.02	0.15	0.40
1,900			0.01					0.01	0.08					0.06			0.10	0.26
2,000			0.01					0.01	0.03	0.03				0.03			0.13	0.24
2,100			0.01						0.03	0.05	0.04			0.02			0.07	0.18
2,200									0.02	0.04	0.02			0.03			0.05	0.18
2,300		0.01	0.00						0.01	0.03	0.09			0.02				0.09
2,400		0.02	0.03						0.03	0.03								0.20
Totals	0.91	0.50	0.62	0.40	0.67	0.40	0.67	0.55	0.68	0.70	0.60	0.11	0.30	0.60	0.29	1.0	1.5	
Wind Speed	3.1					3.6						1.3				2.5		
Wind Direction	10.5%																	

0.0458 = 1 hr, 0.092 = 2 hr, etc.

study area to the conditions observed at the historical meteorological monitoring site (*e.g.*, airport). Since historical data is frequently collected at a site located a considerable distance from the study area (where topographic features may be different from those near the project) there may be some concern regarding the "representativeness" of the historical data to conditions occurring in the vicinity of the project. Through correlation of field measurements taken near the project with similar measurements taken at the historical meteorological site, the "representativeness" of the historical data, when applied to the study area, can be verified.

Once it has been determined that a mesoscale data verification field study is needed, then the following procedure may be followed (see flow diagram in Figure 105):

1. Identify, using a topographic map, possible locations within the study area where general air movements could be measured that would be representative of the study area. Critical locations should be identified using air quality considerations as the criteria. Mesoscale air movements should be measured in areas where there is the highest potential of adverse air quality effects.

2. Select meteorological monitoring equipment. Use rugged, reliable equipment for mesoscale field studies. Relatively slow response instruments are acceptable because the measurement of average conditions is the objective.

3. Deploy instruments at the site or sites selected as representative of the study area. Make sure instruments are exposed properly for a mesoscale site.

4. Conduct study for from one to three months. Operate instruments continuously and record results.

5. Collect and reduce data. Correlate wind speed and direction data from field study with similar measurements taken at the historical meteorological site.

6. Determine if field measurements correlate significantly with historical site. If yes, then no further field study is needed. Historical data can be "corrected" using the results of the correlation, and the air quality impact assessment can be performed using corrected historical data. If the field measurements do not correlate significantly with data from the historical station, then the field study should be continued · for a full year in order to document the entire spectrum of meteorological events that occur during all seasons of the year.

7. Collect and analyze one year's field observations.

8. Perform air quality impact assessment using field sampling data.

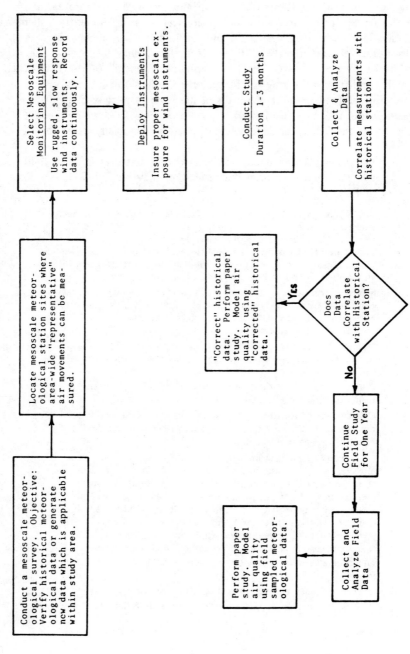

Figure 105. Conducting a mesoscale meteorological survey for data verification.

Meteorological Measurements in
Air Quality Model Validation

Meteorological measurements are also taken in conjunction with air quality surveys designed to validate mathematical air quality simulation models. Both microscale and mesoscale air quality model validation studies can be conducted. The types of meteorological measurements required for model validation purposes depend to a large extent on the technical sophistication of the model to be validated and whether it is a microscale model or a mesoscale model.

When validating a mesoscale model, field measurements of all meteorological parameters in the model are required. Even the simplest mathematical models of air pollution dispersion include the variables, wind speed and wind direction, and some mesoscale models require that the atmospheric stability (turbulent condition) and/or the mixing depth be determined. Photochemical models may require that measurements of the intensity of solar radiation be performed. For validating these more sophisticated models, some method of measuring each parameter in the model will be needed.

In general, the procedure for conducting a meteorological survey for validating a mesoscale mathematical air quality simulation model is as follows (see flow diagram, Figure 106).

1. Identify, using a topographic map, meteorological monitoring locations that are properly exposed to measure "representative" mesoscale air movements near preselected air quality monitoring sites.

2. Select meteorological monitoring equipment to be used during field study. Use rugged, reliable wind speed and direction equipment at sites where only wind speed and direction measurements are needed. At sites where the measurement of the standard deviation of wind direction, σ_θ, is desired, fast response wind vanes are required. At sites where mixing depths are to be measured, lidar, radiosonde, or wiresonde equipment may be needed. If solar radiation measurements are required, then the proper instrumentation must be selected.

3. Conduct study for three to six months. Schedule meteorological measurements to be taken simultaneously with air quality measurements.

4. Collect and reduce data. Use meteorological and air quality measurements to validate mathematical simulation model.

5. Determine if the mathematical model has been adequately validated. If yes, then no further field study is required. If the field measurements do not verify the accuracy of the model, then the field study may be continued for a longer duration, until sufficient data has been collected to validate the model, or to develop an empirical model.

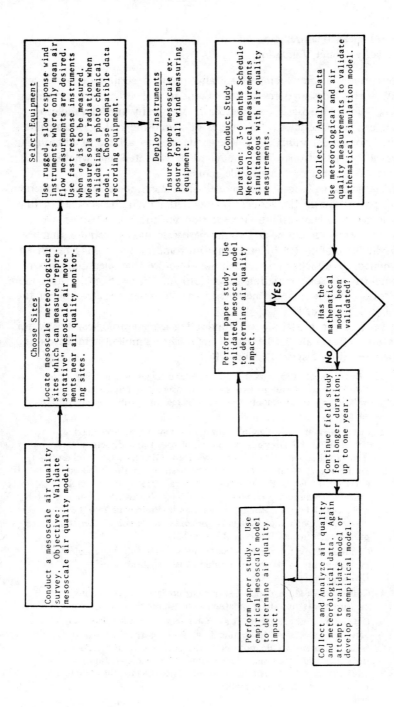

Figure 106. Conducting a meteorological survey in order to validate a mesoscale air quality model.

Validating a Microscale Air Quality Model

A microscale air quality model validation study is conducted when it is expected that microscale terrain features will significantly influence the transport and dispersion of air pollutants generated by the protect in a manner that is unpredictable using an existing simulation model (*i.e.,* where basic assumptions of the model are violated). A microscale air quality monitoring study can then be conducted in order to measure the effects of local terrain or other roughness features on the transport and dispersion of air pollutants within the microscale regime.

As with mesoscale model validation studies, field measurements of all meteorological parameters in the microscale model must be made. In most cases, however, microscale models contain only wind speed, wind direction, and atmospheric stability as important meteorological variables.

Because air movements in the microscale are often highly variable and quite turbulent due to mechanical mixing around surface roughness features, they are usually difficult to describe and quantify from microscale measurements of wind speed and direction. For this reason, the effects of microscale air movements in dispersing air pollutants can probably best be determined by *air quality measurements* rather than microscale meteorological measurements. Air quality measurements taken within the microscale regime, plus simultaneous measurements of mesoscale air movements taken at a nearby meteorological station, provide useful information regarding the effects of mesoscale air movements on microscale dispersion characteristics. If *microscale air quality* can be measured and correlated with local mesoscale meteorology, and if the local mesoscale meteorological station can be correlated to the historical meteorological station, then the microscale air quality condition can be estimated for any mesoscale meteorological condition observed at the historical station, and equally important, the frequency of occurrence of adverse dispersion conditions at the microscale location can then be estimated using data from the historical meteorological station.

In general, the procedure for conducting a meteorological survey for validating a microscale mathematical air quality simulation model is as follows (see flow diagram, Figure 107):

1. Locate one mesoscale meteorological wind speed and direction monitoring station within the same mesoscale regime as the study site where air quality measurements are taken. Sites where microscale air movements can be measured may also be identified.

2. Select meteorological monitoring equipment to be used during field study. Use fast response wind speed and direction instruments at both the microscale and mesoscale meteorological monitoring stations. Fast response wind vanes can then be used

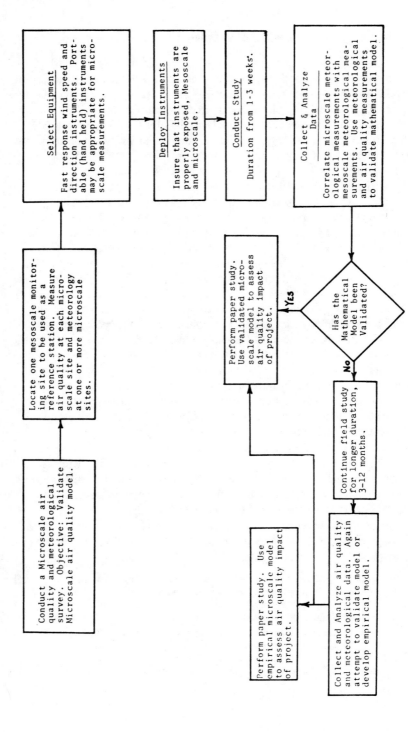

Figure 107. Conducting a meteorological survey in order to validate a microscale air quality model.

to determine the standard deviation of the wind direction, σ_θ, and will provide additional information upon which to estimate atmospheric stability. A permanent type wind instrument should be used at the mesoscale meteorological monitoring site, while portable (hand held, etc.) or permanent type instruments can be used within the microscale regime.

3. Deploy instruments at the site or sites selected for the field study. Insure proper mesoscale exposure for mesoscale site and representative microscale exposures at microscale sites.

4. Conduct study for at least one to three weeks. Schedule meteorological measurements to be taken simultaneously with air quality measurements.

5. Collect and reduce data. Use meteorological and air quality measurements to validate mathematical simulation models.

6. Determine if the mathematical model has been adequately validated. If yes, then no further field study is required. If the field measurements do not tend to verify the accuracy of the model, then the field study may be continued for a longer duration, until either sufficient data has been collected to validate the model or to develop an empirical model.

REFERENCES

1. Beaton, J. L., A. J. Ranzieri and J. B. Skog. "Meteorology and Its Influence on the Dispersion of Pollutants from Highway Line Sources," *Air Quality Manual, Vol. I*, California Department of Public Works, Division of Highways, Sacramento, California (April 1972).

2. Stern, A. C., Ed. *Air Pollution Vol. I*, 2nd Ed. (New York: Academic Press, 1968).

3. Magill, P. L., F. R. Holden and C. Ackley. *Air Pollution Handbook* (New York: McGraw-Hill Co., 1956).

4. Strauss, W. *Air Pollution Control, Part I* (New York: Wiley Interscience, Inc., 1971), pp. 1-35.

5. Slade, D. H., Ed. "Meteorology and Atomic Energy—1968," (Oak Ridge, Tennessee: U.S. Atomic Energy Commission, Office of Information Services, July 1968).

6. *Proc. Symposium on Multiple Source Urban Diffusion Models*, Air Pollution Control Office Publication No. AP-86, Chapel Hill, North Carolina (October 1969).

7. Ludwig, F. L. and W. F. Dabberdt. "Evaluation of the APRAC-1A Urban Diffusion Model for Carbon Monoxide, Final Report CRC/EPA Contract CAPA 3-68 (1-69)," Standard Research Institute, Menlo Park, California, NTIS No. PB 210 86 9 (1972), 147 pp.

CHAPTER XIII

METEOROLOGICAL MEASUREMENTS

INTRODUCTION

Frequently when field measurements of air pollution concentrations are made, it is also desirable to measure the prevailing meteorological condition during the same time period, especially for those projects important enough to require model validation or data verification. When analyzing fluctuations in ambient air pollution levels, the data analyst can often attribute changes in pollution levels to changes in meteorology when meteorological data is available. Without this data, the analyst can only guess at the causes of observed variations in ambient pollution levels.

The most important meteorological parameters affecting the transport and dispersion of air pollutants are wind speed, wind direction and atmospheric stability. Solar insolation plays an important role in the generation of reactive photochemical air pollutants (*i.e.,* ozone and other oxidants). The types of equipment required to measure these important meteorological parameters are discussed in the remainder of this chapter.

WIND SPEED

The surface wind speed can be measured using various types of wind speed devices including pressure plate anemometers, bridled cup anemometers, pressure tube anemometers, and hot wire or heated thermometer anemometers, each described by Middleton *et al.*[1] The most common wind speed measuring instruments, however, are rotating anemometers, such as the propeller or windmill anemometer and the spinning cup anemometer. Figure 108 illustrates different types of rotating anemometers commonly used in air flow studies. Some of the most important characteristics of rotating anemometers are discussed below.

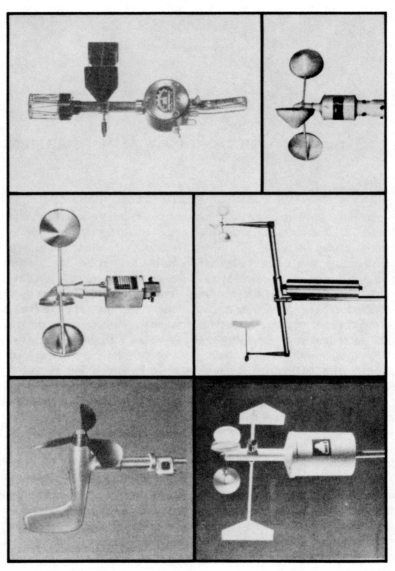

Figure 108. Examples of commercially available wind speed sensors: (1) Bendix Aerovane, (2) Belfort Cup Anemometer, (3) Belfort Hand Held Wind Instrument, (4) Meteorology Research, Inc. Model 1074 Wind Instrument, (5) Climatronics Corp. Wind Mark I, (6) Belfort No. 1259B Cup Anemometer.

Sensitivity

The sensitivity of a rotating anemometer is usually reported as the minimum wind speed that the instrument can accurately measure (starting threshold). Whether or not an anemometer responds to very low wind speeds (*i.e.,* < 1 m/sec) will depend on the mechanical friction and moment of inertia of the rotating vanes of cups. Anemometers designed to be reliable and durable over long periods of operation, such as those used by the U.S. Weather Bureau and the Federal Aviation Agency, must be ruggedly built. These instruments usually employ a cupwheel 38 to 45 cm in diameter from the center of rotation to the center of the cups and utilize conical beaded cups about 10 cm in diameter.[2] The moment of inertia and mechanical friction for this type of anemometer are relatively large, causing the instrument to be fairly insensitive to wind speeds less than 1 m/sec.

For accurate wind measurements at low speeds and for more sensitivity (more rapid response to fluctuations in wind speed), small three-cup anemometers are available. These instruments have cupwheel diameters ranging from 5 to 10 cm and cup diameters ranging from 2 to 6 cm. Cups are made of lightweight durable plastic or aluminum. The moment of inertia and mechanical friction for these small lightweight anemometers are less than for the heavier anemometers, allowing for improved response characteristics.

The vane or cup configuration, the moment of inertia, and the mechanical friction of an anemometer govern the total response characteristics of the instrument. The average amount of torque transmitted by the wind to the anemometer wheel will depend largely on the configuration of the vanes, blades, propeller, or cup. This torque will cause the anemometer wheel to turn as the wind blows against it. The speed of rotation of the cupwheel about the cup centers will generally be between one-half and one-third of the linear velocity of the wind, depending on the mechanical friction, moment of inertia and design configuration of the anemometer.

The sensitivity of the instrument to changes in wind speed (*i.e.,* gustiness) also depends on these three physical parameters. In general, the heavier, more durable anemometers do not respond as well to rapidly fluctuating wind speeds as do the lighter, smaller, more delicate spinning cup and propeller-type anemometers. For this reason, the lighter, more sensitive anemometers are more appropriate for air turbulence studies (microscale monitoring) while the heavier, more durable, anemometers are appropriate whenever average wind speed measurements are needed without measuring gustiness (mesoscale monitoring).

The response rate of an anemometer is commonly measured and reported by the manufacturer as the "distance constant," *i.e.,* the distance that the wind must pass the anemometer before it responds to (1-1/e) or 63% of the change in wind speed. For any anemometer, the distance constant is independent of wind speed. Values of the distance constant range from about 0.7 to 8.0 meters for commercially available anemometers. The smaller the distance constant, the more sensitive the anemometer is to rapid changes in wind speed. Table 20 lists the distance constants for several commercially available anemometers.

Table 20.   Anemometer Characteristics

Manufacturer and Instrument	Model	Distance Constant, m
Teledyne Geotech.:		
Standard 3-cup	Series 50	1.5
Staggered 6	Series 50	1.0
Stainless Steel 3-cup	Series 50	2.4
Bendix Friex Instrument Division:		
Aerovane (3-blade prop.)	Model 120	4.6
Aerovane (6-blade prop.)		5.8
Climet Instruments Inc.:		
3-cup	WS-011	1.5
4-blade prop. (bivane)	018-10,-30	0.9
Electric Speed Indicator Co.:		
3-cup	Type F-420C	7.9
R. M. Young Co.:		
Gill anemometer, 3-cup		0.73
Meteorology Research, Inc.:		
Velocity Vane	1057	1.7 to 1.8
Mechanical Weather Station	1071-1075	5.5
Vector Vane (bivane)	1053 III	1.0
C. W. Thornthwaite Associates:		
3-cup		1.07
Climatronics, Corp.		
3-cup	F460	1.5
3-cup	WM-I	0.76
3-cup	WM-III	2.5
Belfort Instrument Co.:		
Aerovane (3-blade prop.)	Model 120	4.6
3-cup	1250B	7.6
3-cup	Type M	6.4

Sensor Output

The type of sensor output, which varies for different anemometer designs, can be either electrical (variable AC or DC voltages or step function pulses) or mechanical (the turning of clock movements, mechanical counters, or a rotating cam).

In the case of a variable voltage output, the rotating anemometer cupwheel drives a small electrical generator (either AC or DC) which produces a voltage output proportional to the speed of the rotating wheel. Step function pulses are generated by anemometers that use the rotation of the cupwheel to activate switches or relays. The number of pulses per unit time is proportional to the speed of rotation of the anemometer.

Anemometers with mechanical outputs generally have the spinning shaft of the anemometer directly coupled to either a clock-type movement, a mechanical counter, or a rotating cam. The output from the anemometer varies as a function of the number of rotations of the spinning anemometer cupwheel. In effect, mechanical outputs simply count, or accumulate, the total number of revolutions of the cupwheel. The total number of revolutions of the cupwheel per unit time depends on the linear displacement of the wind, or "wind run," that occurs during the time period. Although most anemometers with mechanical outputs measure wind run rather than wind speed, average wind speed (measured in kilometers per hour) can be determined from wind run by dividing the wind run (measured in kilometers) by the time period elapsed during the measurement (hours).

Data Record

The form of the data record generated by the anemometer sensors varies considerably for different types of instrument designs. Wind run measurements from anemometers with mechanical output may be recorded continuously on a mechanical strip chart recorder, or a mechanical counter or clock-face-type meter may be observed periodically and recorded manually. Wind speed measurements from anemometers with electrical outputs can be recorded on standard strip chart recorders or electrical data loggers. Anemometers using a step function electrical pulse output require an "event recorder" which marks on the chart paper the time when each pulse occurs. In some systems, where pulses are very frequent, the signal can be electronically conditioned to record only one pulse per 100 or 1000 that occur, while in other systems the pulses are so rapid that the signal can be electronically conditioned to simulate the output of the variable voltage-type anemometer.

The signal output from variable voltage anemometers is usually recorded on standard potentiometric strip chart recorders. These recorders respond rapidly to changing anemometer outputs and generate a continuous record of instantaneous wind speed with time. This type of wind record contains the most information about the observed condition of the wind. A continuous record of variations in wind speed can be evaluated to determine the mean wind speed, the standard deviation in wind speed, and the maximum and minimum wind speed occurring during the sampling period. In most cases, only the mean wind speed can be determined from event recorders or wind run records. Figure 109 illustrates three types of wind records generated by commercially available equipment.

WIND DIRECTION

The surface horizontal wind direction can be measured using various types of wind direction equipment including one of the oldest and most common meteorological instruments, the wind vane. Fundamentally, a wind vane is a body mounted unsymmetrically about a vertical axis, on which it turns freely. The end offering the greatest resistance to the motion of the wind goes to the leeward. Numerous types of vanes with varying operating characteristics have been constructed. Figure 110 illustrates different wind vane configurations common to commercially available equipment. Some of the most important operating characteristics of wind vanes are discussed below.

Sensitivity

Two important measurements of the sensitivity of a wind vane are the starting threshold, defined as the minimum normal wind speed required to initiate a turn or movement in the wind vane, and the distance constant or natural gust wavelength of the wind vane. As in the case of the distance constant of an anemometer, the distance constant for a wind vane is dependent on the mechanical resistance, the moment of inertia and the design configuration of the instrument. While some manufacturers report the distance constant as the distance the wind has passed the wind vane when it responds to 50% of a step change in wind direction, as suggested by MacCready and Jex,[3] other manufacturers report it as the length of wind trajectory necessary to complete an oscillation of the vane. This measurement, called the natural gust wave length as suggested by Gill,[4] can be divided by [6-2.5 (damping ratio)] to determine constants as measured according to MacCready and Jex's method.

(a)

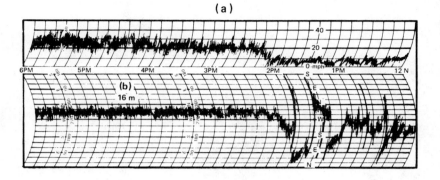

(b)

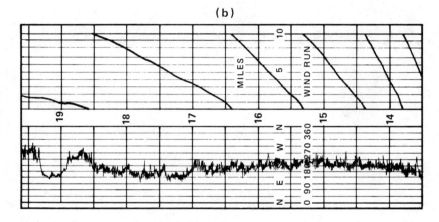

(c)

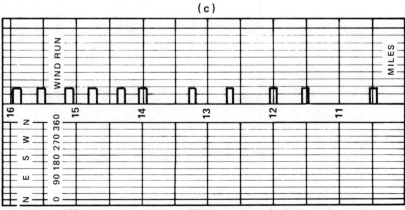

Figure 109. Examples of three different types of wind speed recordings: (a) Continuous, instantaneous wind speed record, (b) Continuous wind run record, and (c) Wind run.

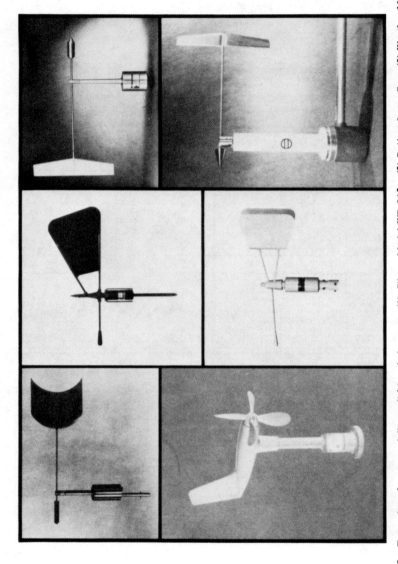

Figure 110. Examples of commercially available wind vanes: (1) Climet Model WD-012, (2) Belfort Instr. Co., (3) Weather Measure Corp. Model W104-2, (4) Weather Measure Corp. Model W101-P, (5) Belfort Instr. Co. Cat. No. 1275C, (6) Teledyne Geotech Model 53.2.

One other common parametric measurement used to describe the response characteristics of a wind vane is the damping ratio. Similar to the drag coefficient for an anemometer, this ratio is a measure of the mechanical resistance to movement of the wind vane. Wind vanes with damping ratios equal to about 0.6 provide the most faithful response to changes in wind direction. Most commercially available wind vanes exhibit damping ratios from 0.1 to 0.7. Instruments with damping ratios greater than one respond very slowly, if at all, to changes in wind direction. Such instruments are referred to as "overdamped" and tend to underestimate the true variability in wind direction. Table 21 lists the damping ratios and gust wavelengths of several commercially available instruments (see Moses[2]).

Table 21. Values of the Damping Ratio and Gust Wavelength for Various Wind Vanes

Manufacturer	Model	Damping Ratio (ζ)	Gust Wavelength (λ_n), m
Bendix Friex Instrument Division:			
Aerovane	120	0.28	10.4[a]
Wind Vane	–	0.5	1.8[a]
Climet Instrument Inc.:			
Wind Vane	WD-012-10	0.4	1.0[a]
Bivane	018-10,-30	0.6	1.0[a]
Electric Speed Indicator Co.:			
Splayed Vane	F420C	0.14	17.7
R. M. Young Company:			
Gill Anemometer-Bivane		0.60	4.4
Belfort Instrument Co.:			
Splayed Vane	Type M	0.20	9.5
Aerovane	Type L & N	0.30	10.4
Meteorology Research, Inc.:			
Wind Vane	1022	0.4	1.13[a]
Vector Vane (bivane)	1053	0.6	1.0[a]
Mechanical Weather Station	1071-1075	0.50-0.60	6.8
Weather Measure Corp.:	W101-P-AC	0.20-0.28	18.0
	W101-P-HF	0.22-0.29	16.5
Climatronics Corp.	F460	0.4	1.12
	WM-I	0.4-0.6	0.76
	WM-III	0.4-0.6	2.45
Teledyne Geotech.	53.1	0.6	1.7[a]
	53.2	0.4	1.1[a]
	107-61	0.2	0.5[a]

[a]Reported as distance constants.

Most properly operating wind vanes measure the mean wind direction accurately, but not all wind vanes respond accurately to rapid changes in wind direction. In general, the larger, heavier, and more durable wind vanes, with characteristically long gust wavelengths, will not respond fully to the rapid changes in wind direction, around one cycle per second, caused by turbulent eddies. The smaller, lighter, and more delicate wind vanes, having short gust wavelengths, respond more accurately to these rapid changes in wind direction. Frequently, in turbulence and air pollution dispersion studies it is desirable to measure the standard deviation in wind direction, σ_θ. If a wind vane used for this purpose does not have a damping ratio close to 0.6 and a relatively short gust wavelength, it will probably underestimate the range of variations in wind direction.

Bivanes

In addition to measurements of the variation in direction of the horizontal wind, wind studies occasionally require measurements of the variation in the vertical wind component. The standard deviation in the direction of the vertical wind component, σ, is an indirect measure of the spectrum of thermal and mechanical turbulence in the lower atmosphere. Several investigators have used field measurements of σ to determine the vertical dispersion parameter, σ_z.[5,6]

Bivanes (*i.e.,* bidirectional wind vanes) are designed specifically for the purpose of measuring both the horizontal and vertical wind directions. Bivanes are very similar in design to standard wind vanes, but they have the ability to swing up and down as well as side to side with changing wind direction. Bivanes are usually made of very lightweight material and use low friction bearings resulting in a very responsive, but delicate instrument. Natural gust wavelengths are generally short. Bivanes are frequently employed in addition to standard wind vanes in atmospheric turbulence and air pollution dispersion studies. Figure 111 illustrates several commercially available bivane instruments.

Signal Output

As with anemometers, the signal output from wind direction sensors can be either mechanical or electrical. The objective of the signal output is simply to transmit to a recording device the position of the wind vane, which is usually measured as degrees from north (*i.e.,* north = 0 or 360°, east = 90°, south = 180°, west = 270°). If the data recording device is located very close to the wind vane, then a mechanical linkage (such as a rotating cam) between the wind vane and the recorder "pen" can be

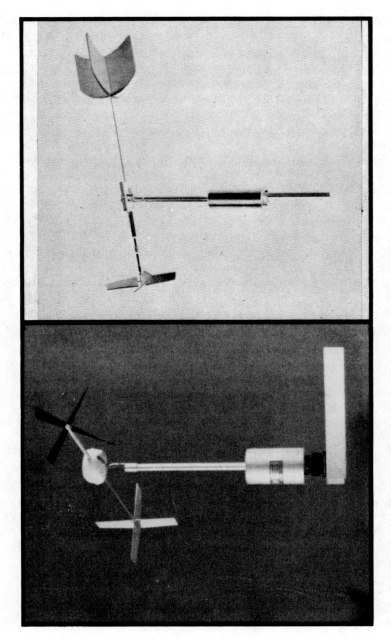

Figure 111. Examples of bivane wind instruments: (1) Meteorology Research Inc., Model 1053, (2) Climet Instrument Co. Model 018-10.

used to record the position of the wind vane. If, as is usually the case, the wind vane is located at an elevated height and the recorder is located at ground level, then some type of electrical signal output may be more desirable.

The most common electrical methods for detecting the position of the wind vane employ either multiple-position switches, position motors, or a slide wire potentiometer, each of which has certain advantages and disadvantages. For example, vanes using position switches are usually cheapest, position motors are the most reliable and maintenance-free, and potentiometers are the most compatible when digitizing or telemetering the output. Special characteristics of each are discussed below.

Wind vanes with position switches generally have a single contact junction fixed to the moving vane shaft and either 4 or 8 separate stationary contacts, wired as separate electrical circuits, positioned in each measurable wind direction. Only one circuit is closed by the rotating contact at any one time corresponding to the position of the wind vane. For example, when the wind vane points north, the circuit wired to the "north" stationary contact is closed while all other circuits are open. The sensor output can be used to light indicator lamps, to trigger a counter or clock, or, with special recorders, to produce a continuous record of wind direction.

Wind vanes employing position motors are very reliable, and, unlike the vanes using position switches, provide good resolution in the final wind direction record. While position switches can resolve the wind direction into only 8 or 16 sectors, position motors resolve the wind direction within about two degrees. Position motors are small motors with a single-phase rotor and three-phase stator (or vice versa). When two or more of these motors are properly connected and supplied with single-phase alternating current, their shafts do not rotate continuously like an ordinary motor, but any rotation which is imposed on one position motor is followed by the others. In a wind vane, the transmitting position motor is connected directly to the wind vane shaft while the receiving position motor can be located elsewhere, either in a console or in a strip chart recorder. The electrical cable connecting the two motors can be quite long as long as certain resistance is not exceeded and the stators are fed from the same electrical supply network. Figure 112 shows a schematic diagram of a pair of position motors that might be used in a wind vane.

The slide wire potentiometer is one of the most common electrical methods used to transmit vane indications in commercial wind vanes. The movable contact of a potentiometer (variable resistor) is fixed to the rotating shaft of the vane. As the vane turns in the wind, the

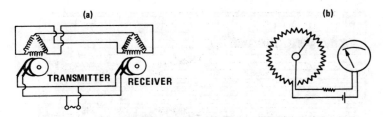

Figure 112. Electrical schematic of typical wind vane sensors:
(a) position motors, (b) potentiometric.

electrical resistance between the two input terminals changes. When a
variable resistor is used in an electrical circuit including a power supply
and a current meter or recorder (see Figure 112), variations in the position or the wind vane will produce similar variations to be observed at
a properly calibrated meter or recorder.

Data Recording

In order to obtain a continuous record of wind direction, the output
of a wind vane is usually recorded on a strip chart recorder. The appear-
ance of the recorder trace depends on the type of sensor output. Wind
vanes employing position switches produce a recorder trace quite different
from vanes using a potentiometer or position motors (see Figure 113).
The wind vane using position switches produces a recorder trace containing a finite number of pen positions, plus lines where the recorder pen
swings from one position to the next, but many of the smaller variations
in position of the wind vane are not sensed or recorded in this type of
unit. Strip chart records of wind vanes using either position motors or
a potentiometer sensor, display all of the variations in position of the
wind vane. One drawback to some of these wind vanes occurs when the
wind direction is recorded using a range of from 0 to 360 degrees.
Since the vane moves 360 degrees in a circle while the recorder moves
linearly, the vane can move back and forth between 1 and 359 degrees,
as frequently occurs when the wind blows from the north. This causes
the recorder pen to swing from one extreme edge of the chart to the
other and results in an ambiguous chart record. The problem is overcome
by using a specially designed vane and recorder employing a 540-degree
chart which eliminates much of the extreme pen swinging, making the
chart record much easier to read (see Figure 114).

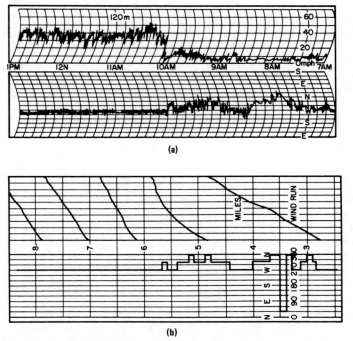

(a)

(b)

Figure 113. Examples of two different types of wind direction traces: (a) from a continuous output wind vane, (b) from a wind vane employing position switches.

ATMOSPHERIC STABILITY

Since the turbulent condition of the atmosphere affects the dispersion of air pollutants, air pollution studies frequently require field measurements of atmospheric stability, which is conventionally reported according to the six "stability categories" suggested by Pasquill.[7] These categories range from extremely unstable conditions, A, to moderately stable conditions, F. Stability category is determined according to the surface wind speed and solar insolation. Table 22 presents the criteria suggested by Pasquill for determining stability categories. These stability categories can be determined by a trained observer in the field by using this table and measuring the wind speed and observing solar intensity, cloud cover and time of day.

Another method sometimes used to estimate atmospheric stability is to measure the standard deviation of the lateral wind direction, σ_θ. This

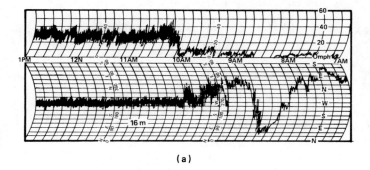

(a)

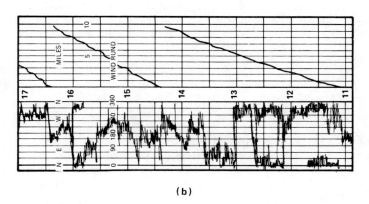

(b)

Figure 114. Examples of two different types of continuous wind direction records: (a) from a 540° wind instrument, and (b) from a 360° wind instrument.

measurement can be made using a fast-response recording wind vane and anemometer erected at the study site. An estimate of the horizontal standard deviation may then be obtained from the wind direction range, that is, the total width of the direction trace over some time interval. According to Slade,[8] the range divided by about 6.0 has been found to give approximation of the horizontal standard deviation for the same data sample period when an averaging time of the order of 1 to 10 sec is used. The σ_θ determined using this method can then be used to estimate Pasquill's atmospheric stability category using Table 23 as presented by Gifford.[9]

Possibly a more appropriate and direct measurement of atmospheric stability is the standard deviation of the vertical wind direction, σ, which

Table 22.   Pasquill's Atmospheric Stability Categories[10]

Surface Wind Speed (m/sec)	Key to Stability Categories				
	Insolation[a]			Night	
				Thinly Overcast or	
	Strong	Moderate	Slight	≥4/8 Low Cloud[a]	≤3/8
< 2	A	A-B [b]	B	–	–
2-3	A-B	B	C	E	F
3-5	B	B-C	C	D	E
5-6	C	C-D	D	D	D
> 6	C	D	D	D	D

[a]Strong insolation corresponds to sunny midday in midsummer in England, slight insolation to similar conditions in midwinter. Night refers to the period from one hour before sunset to one hour after dawn. The neutral category D should also be used, regardless of wind speed, for overcast conditions during day or night, and for any sky conditions during the hour preceding or following night as defined above.

[b]For A-B take average of values for A and B, etc.)

Table 23.   Pasquill's Stability Categories Determined by Standard Deviation of the Lateral Wind Direction

		σ_θ
A	Extremely unstable	$25.0°$
B	Moderately unstable	$20.0°$
C	Slightly unstable	$15.0°$
D	Neutral	$10.0°$
E	Slightly stable	$5.0°$
F	Moderately stable	$2.5°$

can be estimated from the recorder trace of a bivane wind sensor or measured directly using a sigma meter. A sigma meter employs a specially designed electronic circuit which performs electrically the necessary mathematical transformation to convert fluctuating wind directions into a continuous output of the average value of the standard deviation of wind direction, σ_θ or σ. The deviation can then be used to calculate the vertical dispersion coefficient, σ_z, at a distance, x, from the source, using the equation suggested by Islitzer:[5]

$$\sigma x = 1.23 \, \sigma_z \qquad\qquad (13.1)$$

(σ_z and x have units of meters)

An example of a vertical wind direction trace from a bivane wind instrument is shown in Figure 115.

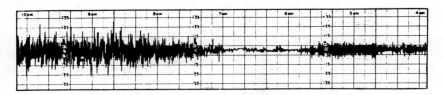

Figure 115.   Example of a vertical wind direction record from a
bivane wind instrument.

With experience, estimates of atmospheric stability can be made by a simple inspection of wind speed and direction records. For example, the wind direction changes very little during stable conditions when there is a prevailing wind but changes abruptly and frequently during low wind speed conditions. During unstable conditions, the wind direction fluctuates continuously, producing a "ragged" recorder trace. Figure 116 illustrates typical wind direction records under four different atmospheric stability conditions reported by Brookhaven National Laboratory.[11]

Another method for determining vertical stability is to measure the vertical temperature gradient of the atmosphere by various techniques, including lidar, wiresonde, acoustic radar, radiosonde, aircraft, and an instrumented tower. Temperature sensors mounted on an instrument tower or an anchored balloon can provide useful data on the vertical temperature gradient very close to the ground. Lidar, radiosonde, acoustic radar and instrumented aircraft are more appropriate when temperature measurements at much greater heights are desired.

Temperature gradients equal to -0.65°C per 100 meters are equivalent to the adiabatic lapse rate and indicate neutral atmospheric stability. This rate of decrease in air temperature with height is normal for the atmosphere under standard conditions. When air temperatures decrease more rapidly with height than the adiabatic lapse rate, the temperature gradient is referred to as superadiabatic, indicating unstable atmospheric conditions. When air temperatures decrease at a lesser rate than adiabatic, as when temperatures are constant or increase with height, a subadiabatic lapse rate and stable atmospheric conditions are indicated. This is the condition commonly referred to as an inversion. Temperature inversions that occur within the lower atmosphere down to ground

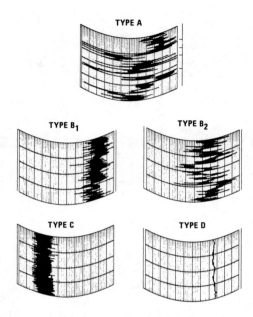

Figure 116. Typical horizontal wind direction traces representing five gustiness classes (From Singer and Smith[11]).

level are referred to as surface-based inversions while those that occur only at elevated heights are generally subsidence inversions. The height measured from ground level to the bottom of a subsidence inversion is referred to as the "mixing depth."

SOLAR RADIATION MEASUREMENTS

Solar radiation measurements may be desirable for two reasons: (a) measurements of solar radiation intensity can be used in conjunction with wind speed measurements to determine atmospheric stability categories, and (b) solar radiation intensity is important in the formation of reactive photochemical air pollutants (*i.e.*, oxidants). Therefore, when photochemical air pollutants are monitored, the intensity of solar radiation during the same time period may be an important and useful measurement. Two types of radiation measuring instruments are commonly used: the pyrheliometer and the net radiometer.

The pyrheliometer measures solar and sky radiation as the radiation heats a small horizontally oriented thermopile enclosed in a glass bulb filled with dry air. An electrical voltage is generated by the instrument in proportion to the intensity of the radiation impinging upon the surface of the thermopile.

The net radiometer measures the net radiation flux across a horizontally oriented flux plate, which is exposed both to sky radiation and ground radiation. The thermopile measures the differential flux between the radiation from the sun and sky and the radiation heat flux from the ground. The entire unit is mounted in front of a housing containing a blower that provides an air flow distributed equally across the top and bottom of the flux plate to equalize convective heat losses and minimize dust and dew accumulation. The net radiometer, however, cannot be used during rainfall since wetting of the top surface causes appreciable cooling and thus produces a spurious indication of radiation flux.

Radiation instruments should be located so that sunlight be unobstructed at all times of the year. The pyrheliometer can be mounted on top of a building if a suitable ground location is unavailable. However, building roofs are not suitable for net radiometers since both the bottom and top plate radiation are of concern. Both types of instruments must be mounted on a firm base and leveled.[2]

EXPOSURE OF WIND INSTRUMENTS

Regardless of the type of wind vane and anemometer used, it is important that the instruments be exposed properly. Wind instruments measure the flow of air that occurs at the point where they are located, and no other. Therefore, if the measurements made by the instrument are to be used to describe the motion of air at some other point (even a few meters from the instrument), then the wind instrument must be exposed to the same air mass without any obstructions interfering with the flow of air. For example, a wind instrument located on the side of a steep hill, near the edge of a cliff, on the side of a building, or too near trees or buildings measures only the *effect of these obstacles on the wind*, not the general movement of air. In some cases, it may be desirable to study the localized influence of terrain or man-made structures, but even so, the general movement of air should also be determined. In most cases, a measure of the general movement of air in the study area is most important, while the localized effects of obstacles on air movement is of secondary importance. The location of the wind instrument should be chosen on the basis of the measurement of objectives. If the objective is to determine the general movement of air within a mesoscale regime, then the instrument should be exposed to air movements that are representative of the mesoscale regime. If, on the other hand, the objective is to study the movement of air within a few meters of the ground, or downwind of certain unusual configurations or terrain features, then measurements of air movements within these microscale regimes may be necessary.

To measure general air movements in the mesoscale, the instrument should definitely be mounted on a mast or meteorological tower. The height of the tower may vary for different applications, but typically, a standard height of 10 meters above ground level is used. The tower or mast should be located in open terrain (representative of the study area), well removed from the influence of roughness features (*i.e.,* hills, trees or buildings). Roughness features generally affect the wind flow within (a) a horizontal distance equal to the height of the roughness feature in the upwind direction, and 5 to 10 times the height of the feature in the downwind direction, and (b) vertical distance equal to the height of the roughness feature (applies mainly above the feature). The exact extent of the region of influence is also dependent on the horizontal extent of the roughness feature. Figure 117 illustrates these "guidelines" for proper exposure of a wind instrument designed to measure general air movement in the mesoscale.[1,2]

In some cases, especially in areas of rough terrain (hills or buildings) or in valley situations, the general air movement within the mesoscale cannot be measured with a single instrument. Since channeling. thermal convection and drainage winds in these areas result in considerable spatial variation in the wind field, no single site can be chosen as representative of the general movement of air. Consequently, measurements at several different sites may be necessary.

When studying microscale air movements close to the ground or near roughness features, wind-measuring equipment may be located within the region of influence of the features under study. The exact location of the wind-measuring equipment will vary depending on the nature of the roughness feature being studied. Frequently, microscale studies require that wind measurements be taken at many different locations in order to determine complex flow patterns. Under these circumstances, portable, hand-held wind-measuring instruments, which can be moved rapidly from place to place, providing measurements from many different locations in a short period of time, can be used.

The primary objective of all microscale air movement studies should be to determine the nature and the extent that the roughness feature influences the general movement of the air mass within the study area. Therefore, whenever wind measurements are taken in the microscale, similar measurements of the general air movement in the mesoscale should also be made. A sufficient number of microscale and mesoscale

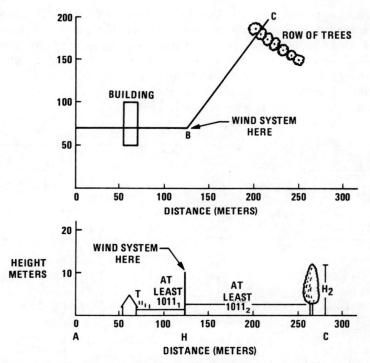

Figure 117. Proper exposure of a wind instrument to measure the mesoscale surface wind.[12]

measurements should be made so that under critical examination the microscale measurements can be separated into two components based on the general air flow in the area and the localized influence of roughness features.

Two common mistakes in wind instrument siting and data interpretation deserve additional comment. One is the placement of wind instruments intended to measure the mesoscale wind on top of buildings, and the other is ignorance of wind shear in microscale studies. Both stem from a lack of understanding of the influence of roughness features on air movement. The installation of a wind instrument above a large building practically insures that it will record a fictitiously high speed because the streamlines will be crowded together as the wind passes the building. The gustiness of the wind will be exaggerated; the measured wind direction will vary widely as the wind is deflected off various structural features of the building. Nearby buildings may also affect the measurement. The resulting wind measurements will be representative of neither general mesoscale air movement nor of conditions at ground level. Wind instruments can be effectively located on buildings, however, if the equipment is mounted on a mast or tower tall enough to place the wind instrument outside the region of influence of the building, usually 10 meters or more above the roof. The effect of a sharp-edged building on the vertical wind profile is illustrated in Figure 118.

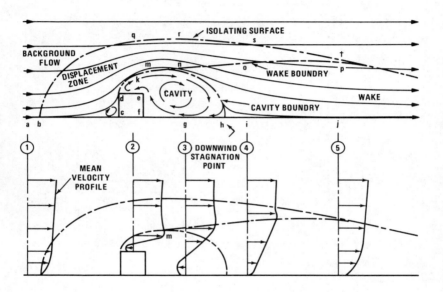

Figure 118. Effect of a sharp-edged building on the verticle profile of horizontal wind speeds (From Slade[8]).

In microscale studies, especially model validation studies, air pollution concentrations downwind of a highway source are usually measured at a height of 1.575 m (5 ft) while wind speeds may be measured only at 10 m or higher. When this is done, the wind speed measured at 10 m will generally overestimate the wind speed occurring much closer to the ground at 1.575 m, where the surface of the earth slows down the movement of air (see Figure 119).[13] Geiger[10] reports the findings of C. W. Thornthwaite who measured variations on the order of 30% between the wind speed measured at 1.5 m and 10 m. Geiger reports that the simplest form of the equation relating wind speed, u, and height above the ground, z, is:

$$u = u_1 \, z^a \qquad\qquad (13.2)$$

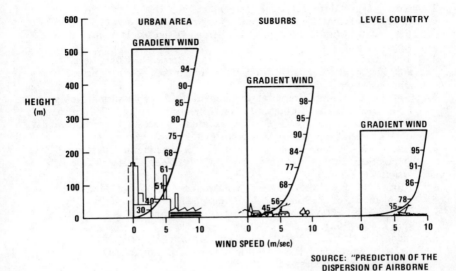

Figure 119. Effects of terrain roughness on the vertical profile of horizontal wind speeds (From Beaton *et al.*[13]).

where: u_1 = wind speed at 1 m
        a = an empirically derived coefficient.

Values of "a" depend on the temperature structure and the roughness of the ground surface. Geiger reports "a" values ranging from 0.07 during daytime in the summer to 0.59 for nighttime in the winter. If wind speeds do indeed vary greatly between 10 m and 1.5 m above the ground, then the accurate measurement of the wind speed responsible for the transport and dilution of air pollutants becomes considerably more complex. If a strong vertical gradient in the wind speed near the ground is expected, it may be necessary to measure the wind speed at more than one height. With more than one measurement of wind speed, an empirical value for "a" in Equation 13.1 can be determined and used to describe the vertical wind profile.

REFERENCES

1. Middleton, W., E. Knowles and A. F. Spilhaus. *Meteorological Instruments*, 3rd ed. (Toronto: University of Toronto Press, 1953).

2. Moses, H. "Meteorological Instruments for Use in the Atomic Energy Industry," in *Meteorology and Atomic Energy*, D. H. Slade, Ed. (U.S. Atomic Energy Commission, Office of Information Services, July 1968).

3. MacCready, P. B., Jr. and H. R. Jex. "Response Characteristics and Meteorological Utilization of Propeller and Vane Sensors," *J. Appl. Meteorol.* 3(2):182-193.

4. Gill, G. C. "Meteorological Instrumentation, Professional Notes, C. E. 199, Partial Bibliography on Anemometers and Calibration Devices," lecture by G. C. Gill, assisted by Floyd Elder, University of Michigan, Ann Arbor, Michigan (1959).

5. Islitzer, N. F. and R. K. Dumbauld. "Atmospheric Diffusion-Deposition Studies over Flat Terrain," *Internat. J. Air Water Pollution* 7(11-12):999-1022 (1963).

6. Islitzer, N. F. "Short Range Atmospheric Dispersion Measurements from an Elevated Source," *J. Meteorol.* 18(4):443-450 (1961).

7. Pasquill, F. *Atmospheric Diffusion*. (London: D. Van Nostrand Co. Ltd., 1962).

8. Slade, D. H., Ed. *Meteorology and Atomic Energy, 1968*. (Oak Ridge, Tennessee: U.S. Atomic Energy Commission, Office of Information Services, July 1968).

9. Gifford, F. A. "An Outline of Theories of Diffusion in the Lower Layers of the Atmosphere," in *Meteorology and Atomic Energy, 1968*, D. H. Slade, Ed. (Oak Ridge, Tennessee: U.S. Atomic Energy Commission, Office of Information Services, July 1968), pp. 65-116.

10. Geiger, R. *The Climate Near the Ground*. (Cambridge, Massachusetts: Harvard University Press, 1965).

11. Singer, I. A. and M. E. Smith. "Atmospheric Dispersion at Brookhaven National Laboratory," *Internat. J. Air and Water Pollution*, 10:125-135 (1966).

12. *Meteorological Aspects of Air Pollution*, Training Course Manual Public Health Service, Division of Air Pollution, HEW (April 1968).

13. Beaton, J. L., A. J. Ranzieri and J. B. Skog. "Meteorology and Its Influence on the Dispersion of Pollutants from Highway Line Sources," in *Air Quality Manual, Vol. I* (Sacramento, California: California Department of Public Works, Division of Highways, April 1972).

CHAPTER XIV

AIR QUALITY DATA SUMMARIES
AND DATA PRESENTATIONS

Various types of data summaries and presentation methods, including graphic illustrations of air pollutant trends, isopleth mapping of pollutant concentrations, and tabular methods of presenting air quality data, may be the best method of illustrating or describing the temporal and spatial resolution of air pollution concentrations as determined from field measurements.

TABULAR DATA SUMMARIES

All types of air quality and meteorological data can be presented in tabular data summaries. Tabular summaries are versatile and relatively easy to prepare. They can be used to compare the pollution concentrations measured at different locations, on different days, at different times of day, for different pollutants, at different frequencies of occurrence of high concentrations or at different averaging times. Tables 24-27 illustrate some examples of tabular air pollution data summaries that may be appropriately used in air quality impact reports.

PRESENTATION OF TRENDS

Most air pollution concentrations vary considerably from hour to hour, day to day, season to season and year to year. Furthermore, these variations are usually not random but follow fairly predictable temporal patterns according to the season of the year, day of week and hour of the day. Predictable variations are usually referred to as air pollution trends.

Trends in air pollution concentrations are probably best illustrated using graphs, which can illustrate diurnal, daily, seasonal or yearly

Table 24. Air Quality Data Summary—Particulate Matter
($\mu g/m^3$ by the high volume air sampling method, June, July, August 1971)

Month, Basin, Station								Days of the Month																							No. ≥ 100 $\mu g/m^3$
	1	2	3	4	5	6	7	8	9	10	11	12	13	14	15	16	17	18	19	20	21	22	23	24	25	26	27	28	29	30	31	
August, 1971	*						*	*						*	*						*	*						*	*			
Northeast Plateau Air Basin																																
Yreka				57					51					35				66						54					60			—
Sacramento Valley Air Basin																																
Chico				73					101					74				101					90					63				2
Sacramento 10th and P				60					88					69				64					78					40				—
Redding Market St.				96					44					69				105					92					74				1
Redding Hospital Ln									81																155								1
Yuba City				89					116					84				115					116					95				3
San Jaoquin Valley Air Basin																																
Fresno				94					127					100									133						71			3
Bakersfield Gld. State Hwy.				168					162																							2
Bakersfield Flower St.				172					156				144					169						136				137				6
Bakersfield Chester Ave.				143					144				108					124					65					116				5
Kern Refuge		96							107				90					8					423						155			3
Taft									83					96				88					94						71			—
Stockton				75					89					70		96							88						55			
Hanford				174					276				181		74			192					200						55			5
Avenal				84					89															78								
Modesto				72					105				96										116						55			2
Visalia				137					126				96										182					161				4
Southeast Desert Air Basin																																
Ridgecrest																								61					62			—
Mojave									91					113				146					173					155			4
Indio			96																													

*Indicates weekends and holidays

Table 25. Air Quality Summary—Oxidant

Maximum Hourly Average Concentration in ppm on Days when a 0.08 ppm Value was Equalled or Exceeded, by Day and Month
June, July, August 1971

Month and Station	Day of Month																															Total No. Exceeded	
	1	2	3	4	5	6	7	8	9	10	11	12	13	14	15	16	17	18	19	20	21	22	23	24	25	26	27	28	29	30	31	Hours	Days
July, 1971	*	*	*	*	*	*				*	*			*		*	*	*					*	*	*						*		
South Coast Air Basin																																	
Ontario	0.12	0.15	0.13	0.12	0.13	0.13	0.13	0.21	0.23	0.26	0.18	0.11	0.24	0.16	0.24			0.18			0.18	0.15	0.13	0.13	0.13	0.25	0.22	0.15	0.15	0.09	0.12	190	26
Redlands	0.10	0.20	0.18	0.14	0.16	0.17	0.21	0.25	0.23	0.31	0.21	0.17	0.26	0.21	0.21	0.16		0.18	0.20	0.20	0.23	0.22	0.20	0.18	0.16	0.23	0.21	0.20	0.17	0.13	0.22	285	31
Santa Barbara								0.08		0.08				0.08																0.10		88	5
Ojai	0.12	0.18	0.16	0.16	0.15	0.19	0.20	0.20	0.18	0.10	0.10	0.18	0.08		0.09		0.08	0.09		0.08	0.10	0.08	0.10	0.09	0.13	0.12			0.17	0.12	0.09	126	10
Camarillo	0.12	0.12	0.15	0.12	0.08	0.10	0.10	0.12	0.15	0.11	0.17	0.11	0.12	0.21	0.21	0.15	0.10	0.26	0.25	0.30	0.30	0.10	0.10	0.21	0.22	0.11	0.13		0.15	0.12	0.09	114	26
Los Angeles	0.08	0.10	0.14	0.22	0.13	0.11	0.14	0.13	0.15	0.15	0.38	0.15	0.25	0.17	0.11	0.14	0.17	0.12	0.12	0.18	0.30	0.17	0.17	0.21	0.15	0.21	0.20	0.19	0.23	0.23	0.28	271	25
Azusa	0.12	0.14	0.14	0.12	0.12	0.20	0.18	0.36	0.36	0.16	0.30	0.15	0.25	0.17	0.08	0.14	0.08	0.12	0.12	0.18	0.17	0.11	0.10	0.11	0.10	0.21	0.20	0.20	0.21	0.21	0.19	211	31
Burbank		0.09	0.10	0.10	0.09	0.10		0.08		0.14	0.13	0.09	0.09	0.08			0.08								0.10	0.09	0.08		0.14	0.13		59	31
West Los Angeles	0.16	0.19	0.12	0.14	0.13	0.14	0.18	0.24	0.17	0.21	0.13	0.13	0.17	0.22	0.18	0.12	0.17	0.15	0.13	0.18	0.18	0.18	0.16	0.17	0.17	0.22	0.22	0.21	0.12	0.16	0.15	8	20
Long Beach	0.13	0.16	0.15	0.16	0.15	0.15	0.19	0.24	0.24	0.34	0.22	0.11	0.30		0.10		0.19	0.22		0.24	0.23	0.18	0.16	0.17	0.17	0.25	0.22	0.17	0.14	0.12	0.21	220	3
Reseda	0.15	0.21	0.22	0.20	0.16	0.08		0.27	0.33	0.33	0.32	0.18	0.33	0.18	0.15	0.13	0.21	0.11	0.11	0.23	0.20	0.20	0.09	0.17	0.21	0.25	0.26	0.15	0.09	0.08	0.21	211	29
Pomona	0.08	0.09	0.10	0.08	0.08	0.08	0.18	0.21		0.21	0.26		0.22	0.22		0.16	0.11	0.09	0.11	0.09			0.09	0.09	0.15	0.15			0.13	0.13	0.22	22	10
Lennox	0.20	0.22	0.13	0.16	0.16	0.15	0.17	0.18	0.23	0.13	0.13	0.16	0.21	0.25	0.23	0.15	0.16	0.12	0.19	0.24	0.23	0.28	0.22	0.18	0.18	0.21	0.30	0.27	0.18	0.19	0.15	263	31
Pasadena																																110	25
Whittier																																257	30
San Diego Air Basin																																	
San Diego		0.08	0.08		0.08			0.08	0.09	0.12																						2	1
Chollas Heights	0.09	0.08	0.09	0.08	0.08		0.10	0.18	0.25	0.13	0.08							0.20	0.18							0.10				0.08	0.11	129	8
El Cajon	0.09	0.08	0.08	0.08	0.09		0.12	0.18		0.27	0.18		0.13	0.10	0.15	0.10	0.18	0.20	0.18								0.08			0.08		102	20
Nestor	0.09	0.10	0.08	0.10	0.09		0.08	0.12	0.18	0.14	0.16	0.09	0.08	0.08	0.08		0.06					0.08				0.08			0.08	0.09	0.09		7
Mission Valley								0.09	0.08	0.13																							8
Oceanside																																	
Sacramento Valley Air Basin																																	
Chico	0.11	0.08	0.09	0.09	0.08	0.10	0.11	0.11							0.09	0.12	0.11	0.12	0.10	0.12	0.11	0.13	0.10	0.12	0.11	0.12	0.13	0.14	0.12	0.12	0.14	241	26
Sacramento—13th & J Sts.	0.09		0.08	0.08	0.08	0.11									0.08	0.08			0.10	0.08	0.10	0.08	0.08	0.09	0.08	0.12	0.13	0.09	0.09	0.09	0.09	19	7
Sacramento—Creekside	0.13	0.10	0.11	0.11	0.10	0.11	0.15							0.10	0.12	0.15	0.09	0.13	0.10	0.08	0.11	0.11	0.10	0.11	0.11	0.12	0.11	0.12	0.09	0.11	0.11	113	23
Sacramento—1000 P St.	0.12	0.13	0.12	0.14	0.10	0.18		0.12	0.08		0.11	0.08	0.08	0.10	0.15	0.16	0.12	0.15	0.12	0.10	0.10	0.10	0.11	0.14	0.15	0.14	0.13	0.12	0.16	0.14	0.15	127	25
Redding		0.08																			0.12	0.16	0.12	0.11	0.10	0.14	0.13	0.14	0.11	0.11	0.15	201	23
Yuba City																						0.11	0.12	0.11	0.11	0.11	0.11	0.12	0.11	0.11	0.11	95	12
San Joaquin Valley Air Basin																																	
Fresno	0.08	0.08	0.09	0.09	0.08	0.09	0.09	0.08	0.09		0.09	0.10	0.17	0.18	0.14	0.15	0.09	0.12	0.15	0.17	0.12	0.09	0.09	0.10	0.10	0.14	0.17	0.14	0.11	0.11	0.09	228	28
Bakersfield—Golden State	0.08		0.08	0.08	0.08								0.08	0.20	0.20	0.08	0.08	0.14	0.15	0.09		0.08	0.09	0.08	0.08	0.08	0.08	0.09	0.10	0.09	0.08	35	15
Bakersfield—Chester	0.12	0.13	0.11	0.14	0.13	0.12	0.13		0.10		0.15	0.15	0.18	0.20	0.17	0.17	0.14	0.14	0.13	0.13	0.10	0.12	0.09	0.12	0.12	0.13	0.14	0.14	0.14	0.14	0.13	240	27
Stockton	0.16	0.11	0.15	0.13	0.16	0.11	0.14			0.09			0.16	0.09	0.09	0.16	0.08	0.14	0.11	0.16	0.16	0.15	0.11	0.11	0.13	0.14	0.13	0.13	0.10	0.13	0.08	29	7
Modesto	0.12	0.11	0.11	0.11	0.10	0.10	0.10	0.09	0.08	0.09	0.09	0.12	0.10	0.12	0.10	0.12	0.09	0.11	0.11	0.10	0.12	0.10	0.10	0.11	0.11	0.13	0.15	0.13	0.12	0.12	0.10	195	28
Visalia																																218	29
Southeast Desert Air Basin																																	
Banning	0.12	0.10			0.08			0.13	0.11	0.16	0.10	0.10	0.15		0.09			0.11	0.15	0.12	0.14	0.10	0.10		0.10	0.13	0.13	0.13	0.11		0.08	0.14	130	23
Palm Springs	0.21	0.18	0.14	0.13	0.08	0.13	0.16	0.17	0.14	0.21	0.12	0.09	0.16	0.20	0.11		0.14	0.16	0.15	0.11	0.10	0.10			0.13	0.15	0.18	0.12	0.10	0.12	0.18	220	18
Indio	0.24	0.20	0.12	0.13	0.16	0.12	0.14	0.21	0.11	0.30	0.16	0.22	0.18	0.22	0.09	0.09	0.08	0.14	0.21	0.08	0.24	0.17	0.20	0.16	0.09	0.09	0.10	0.10	0.09	0.09	0.10	446	28
Victorville	0.12	0.13	0.12	0.10	0.10	0.08	0.10	0.09	0.11	0.11	0.10	0.10	0.16	0.09	0.10	0.10	0.09	0.08	0.10	0.08	0.09	0.10	0.10	0.11	0.13	0.11	0.11	0.12	0.13	0.09	0.10	122	25
Lancaster	0.10	0.13	0.08	0.11	0.09	0.12	0.11		0.11	0.10	0.10	0.10	0.10	0.14	0.15	0.08	0.08	0.09	0.11	0.11	0.10			0.12	0.13	0.11	0.12	0.11	0.08			116	24

Table 26. Air Quality Data Summary—Hi-Vol. Received by the Air Resources Board for 1971

Basin, Station	No. of Samples	No. of Samples Exceeding			Max µg/m³	Min µg/m³	Annual Geometric Mean µg/m³	σ_g
		260 µg/m³	150 µg/m³	100 µg/m³				
North Central Coast Air Basin								
Gonzales	24	—	2	6	211	22	76	1.7
King City	24	—	1	4	156	10	61	1.78
Monterey	25	—	—	—	88	14	41	1.48
Salinas	65	—	3	16	207	—	65	1.6
Santa Cruz	25	—	—	—	74	23	48	1.31
South Central Coast Air Basin								
San Luis Obispo	73	—	—	—	92	13	43	1.51
Santa Maria	75	—	8	24	232	13	80	1.67
South Coast Air Basin								
Orange Co. Airport	37	—	3	13	215	26	87	1.58
Anaheim	88	1	12	32	294	24	88	1.63
Azusa	71	6	37	55	430	23	142	1.69
Camarillo	30	—	2	14	206	32	96	1.49
La Habra	82	5	16	55	365	35	114	1.58
Lennox	71	3	28	65	283	40	139	1.41
Los Angeles	70	2	40	60	309	28	144	1.53
Los Alamitos	35	2	3	13	337	39	87	1.67
Ojai	41	—	1	8	159	21	73	1.52
Oxnard	26	1	1	6	249	36	74	1.60
Pasadena	71	—	6	38	240	24	100	1.46
Riverside	34	5	19	26	609	42	147	1.79
San Bernardino	33	2	18	22	419	28	130	1.83
Santa Barbara	61	—	1	7	168	11	60	1.61
Ventura	36	—	1	6	152	13	73	1.51
San Diego Air Basin								
San Diego	42	—	—	10	140	41	74	1.2
El Cajon	45	—	6	16	210	34	89	1.2

Table 27. Air Quality Data Summary–High Oxidant
Occurrences of Oxidant having a Value Greater than 0.08 ppm for the Year 1971
South Coast Air Basin

Name	Jan Hr	Jan Dy	Feb Hr	Feb Dy	Mar Hr	Mar Dy	Apr Hr	Apr Dy	May Hr	May Dy	Jun Hr	Jun Dy	Jul Hr	Jul Dy	Aug Hr	Aug Dy	Sep Hr	Sep Dy	Oct Hr	Oct Dy	Nov Hr	Nov Dy	Dec Hr	Dec Dy	Annual Hrs	Annual Dys
Anaheim	17	4	7	4	15	6	8	3	16	4	28	9	33	10	29	13	40	12	29	7	4	2	0	0	226	74
Azusa	26	8	45	11	50	13	77	16	82	17	184	25	249	31	233	31	161	25	76	16	44	12	0	0	1227	205
Burbank	11	5	19	7	28	10	42	12	38	11	142	22	185	31	182	31	114	23	41	10	6	3	0	0	808	165
Corona	–	6	–	10	69	17	63	15	52	12	12	5	127	26	129	28	152	24	87	18	23	7	0	0	714	168
La Habra	13	5	15	6	26	11	20	8	15	8	58	15	88	24	103	23	105	22	68	14	21	8	4	2	536	146
Lennox	17	7	19	6	4	1	15	7	0	0	8	4	9	5	3	1	6	3	19	7	0	0	0	0	100	41
Long Beach	6	4	5	4	2	2	12	6	3	1	3	2	7	3	10	7	17	5	13	5	0	0	0	0	78	39
Los Angeles (Downtown)	8	3	16	5	12	4	44	10	16	8	65	16	83	23	70	21	46	13	31	13	2	2	0	0	393	118
Newhall	4	1	2	1	32	7	40	11	88	12	181	25	232	30	155	27	83	18	15	5	17	7	0	0	829	144
Ojai	6	2	18	4	44	10	95	20	163	18	178	23	81	10	152	20	189	24	17	4	23	7	0	0	966	142
Ontario	5	2	25	8	11	6	21	6	15	4	110	18	167	26	82	19	46	16	20	6	0	0	0	0	502	111
Pasadena	12	8	44	12	44	12	69	16	64	16	172	24	239	31	203	31	146	24	90	17	41	13	0	0	1135	204
Pomona	9	4	12	4	33	12	35	10	52	12	109	18	188	29	154	27	107	21	40	9	21	8	0	0	760	154
Redlands	1	1	14	6	54	14	30	7	69	11	145	18	256	31	200	30	142	22	17	4	14	5	0	0	942	149
Reseda	7	3	18	7	42	9	46	11	87	14	199	26	193	31	150	29	103	22	45	8	18	7	0	0	908	167
Riverside	35	10	58	14	105	18	84	18	107	16	234	26	267	31	277	31	236	30	84	15	39	11	0	0	1526	220
San Bernardino	0	0	5	2	26	9	44	11	78	14	213	26	255	31	225	31	142	24	35	8	2	2	0	0	1025	158
Santa Ana	43	9	56	13	34	8	43	13	16	5	23	7	28	8	9	7	48	13	45	10	15	6	0	0	360	99
West Los Angeles	17	5	15	5	10	4	21	9	10	5	28	8	36	16	45	17	25	12	25	9	3	3	0	0	235	93
Whittier	19	7	28	9	12	5	18	6	20	6	46	10	81	21	94	22	87	19	67	15	5	4	0	0	477	124
Total Hours	267		421		653		827		971		2138		2904		2505		1995		864		298		4		13,747	
Total Days		88		128		178		215		194		327		448		446		372		200		107		2		2,861

pollution comparisons. Examples of different ways trends can be presented and compared are shown in Figures 120-126. The comparisons illustrated in each figure are described below:

> Figure 120 illustrates the observed diurnal trend of one pollutant, at one monitoring location, during different seasons of the year.

> Figure 121 illustrates the observed diurnal trend of one pollutant, at one location, compared to the diurnal trend of traffic volume.

> Figure 122 illustrates the observed diurnal trend of one pollutant, at one location, compared for different days of the week.

> Figure 123 compares the observed diurnal trend for four different pollutants measured on the same day, at the same monitoring station.

> Figure 124 illustrates the observed trend in annual air pollution levels for one pollutant over several different years.

> Figure 125 illustrates the diurnal trend of one pollutant observed at different locations on the same day.

> Figure 126 illustrates the seasonal trend of one pollutant observed at different locations over a period of years.

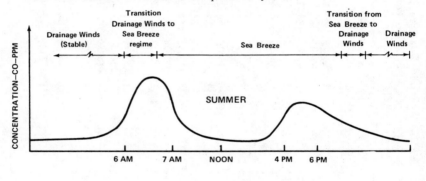

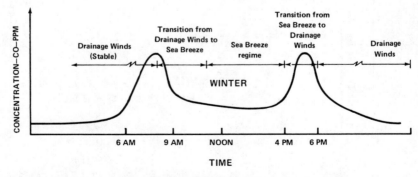

Figure 120. Typical diurnal variation of ground level concentrations of CO for a summer and a winter day.

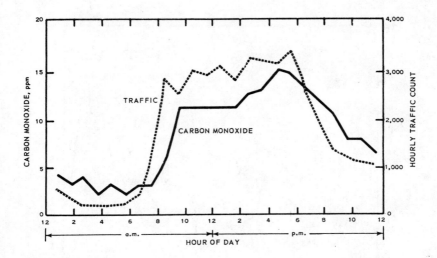

Figure 121. Hourly average carbon monoxide concentration and traffic count in midtown Manhattan.

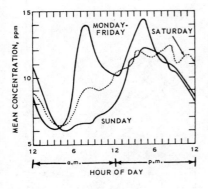

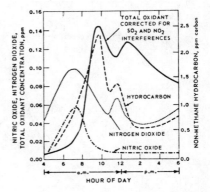

Figure 122. Diurnal variation of carbon monoxide levels on weekdays, Saturdays and Sundays in Chicago, 1962-1964.

Figure 123. Hourly variation of selected pollutants in Philadelphia on Tuesday, July 18, 1967.

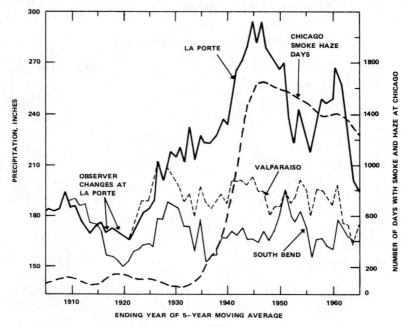

Figure 124. Precipitation values at selected Indiana stations and smoke-haze at Chicago. This figure shows the way in which precipitation trends at La Porte follow the haze changes in Chicago. The results are plotted as five-year moving averages.

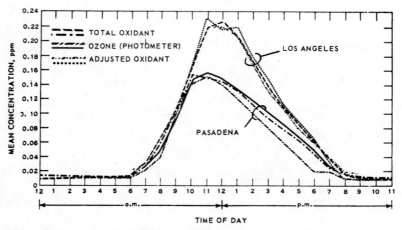

Figure 125. Comparison of the hourly variation of mean 1-hr average concentrations of ozone, oxidant and oxidant adjusted for NO_2 and SO_2 response. Los Angeles and Pasadena, July 1964.

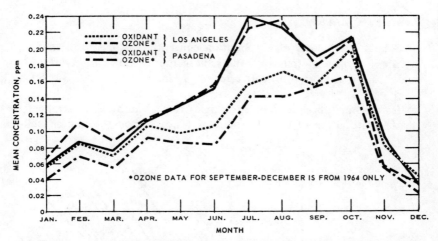

Figure 126. Comparison of the monthly variation of mean daily maximum 1-hr average ozone and oxidant concentrations in Los Angeles and Pasadena, 1964-1965.

Another informative graphic presentation is illustrated in Figure 127. This graph illustrates the typical diurnal trend of 1-hr average CO levels and the peak 1-hr and 8-hr average values that can be compared to the NAAQS. In addition to indicating the highest concentrations observed, it also shows when and what time of day these peak values occurred.

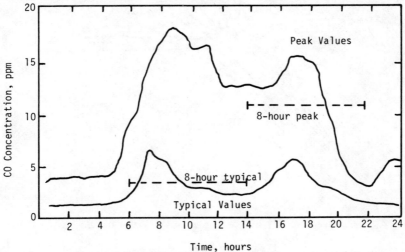

Figure 127. Example plot of typical and peak CO concentrations— continuous and 8-hr average values.

ISOPLETH MAPS

Spatial distribution of air pollution concentrations can be illustrated by a comparison of trends on the same graph as shown in Figures 125 and 126, but a more informative technique may be to use an isopleth map.

Isopleth pollution concentration maps can be prepared illustrating the spatial distribution of average pollutant concentrations, maximum pollutant

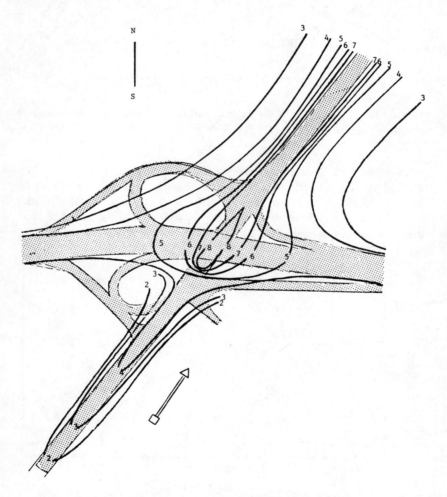

Figure 128. Carbon monoxide isoconcentrations (ppm) for worst possible meteorological conditions at Stratford Road-I-40 Intersection, 1975.

concentrations, typical values during a particular time of day or season of year, or the pollutant concentration distribution that typically occurs under specific meteorological conditions (wind speed, wind direction and stability). Isopleth maps are especially useful for illustrating the size of the geographical area affected by air pollution. Examples are illustrated in Figures 128 and 129.

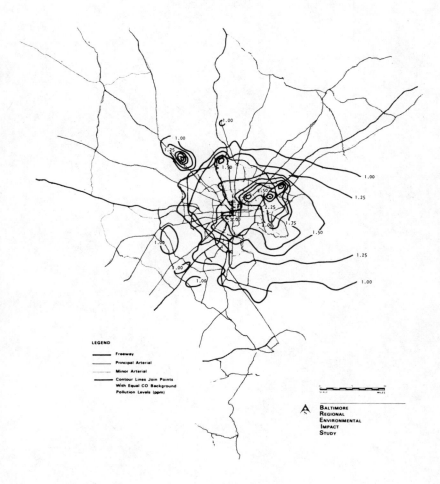

Figure 129. Predicted urban background concentrations of carbon monoxide (ppm) for Alternative 1—existing, 1970.

CHAPTER XV

AIR QUALITY DATA EVALUATION

CALCULATING POPULATION STATISTICS

The collection of data on air quality involves the taking of a limited number of samples from a very variable and uncontrolled population (the environment). Air quality data is analyzed using statistical methods to describe the behavior of the total population based on a finite number of samples. In particular, statistical parameters can be calculated to describe typical and maximum values and the range of data. The first two sections discuss the treatment of typical and peak values; the third discusses the range of the data.

Typical Values

The arithmetic mean, the median and the geometric mean are indicators of typical values. While all three indicate typical values, if the purpose of the summary is to compare the data to the National Ambient Air Quality Standards, then the standard suggests the appropriate statistic. The equations to calculate these parameters are:

Arithmetic Mean:

Given a set of n observations of concentrations, X, *i.e.*, $X_1, X_2, ..., X_n$, the arithmetic mean is

$$\bar{X} = \frac{1}{n} \sum_{i=1}^{n} X_i$$

When the term "average" is used the arithmetic mean is usually implied.

Median:

The median is the middle value of the data; that is, the value that has half the data above and half below. If the data is ranked in order of

251

magnitude so that $X_1 \leqslant X_2 \ldots \leqslant X_n$, the median is either $X_{\frac{n+1}{2}}$ if n is odd, or $\frac{X_{\frac{n}{2}} + X_{\frac{n+1}{2}}}{2}$ if n is even.

The median is a convenient statistic that is not influenced by changes in the extremely high or low values of the distributions, as would be the arithmetic mean.

Geometric Mean:

Given a set of n observations, *i.e.*, X_1, X_2, ..., X_n, the geometric mean, $M_g = (X_1 X_2 \ldots X_n)^{1/n}$. Since this is the least intuitive of the statistics presented, it is worthwhile to discuss it in more detail.

If a distribution is symmetric, such as the normal distribution, the expected value of the arithmetic mean and median are identical. However, for a lognormally distributed variable, it is the expected value of the geometric mean that approximates the expected value of the median. Therefore, since air pollutants often have a distribution that is approximately lognormal (see section on "Frequency Analysis"), the geometric mean is commonly used as a convenient method of summarizing the data; for total suspended particulate matter, for example, the annual standard is based on the annual geometric mean concentration.

As an alternate computational formula, it should be noted that

$$\log M_g = \frac{1}{n} \sum_{i=1}^{n} \log X_i \quad \text{or} \quad M_g = \text{EXP} \frac{1}{n} \sum_{i=1}^{n} \log X_i$$

Maximum Values

Maximum values may be indicated by listing the maximum and/or the second highest value. The second highest value is important because compliance with short-term air quality standards is determined by this value. The principal difficulty in using the second highest value is that it does not allow for differences in sample sizes. For example, if two monitoring devices are side by side, one operating every day of the year and the other operating only every sixth day, it would be expected that the second highest value for the every-day sampler would be higher than that for the other, even though they both monitored the same air. Table 28 illustrates how the second highest value may vary depending upon different sampling frequencies. Table 28 is based upon total suspended particulate data from a Philadelphia site that sampled daily.[1]

To allow for dependence upon sample size, various percentiles are sometimes used to indicate maximum values. For example, the 99th percentile might be used for hourly data, while the 90th might be appropriate for daily measurements. By using a percentile value rather than an absolute count of samples, allowance is made for sampling frequencies

Table 28. Maximum and Second Highest Values (Philadelphia, 1969)
for Various Sampling Schemes

Sampling Schedule	Observations	Maximum	Second Highest
Every day	365	325	244
Every sixth day	61	219	215
	61	195	171
	61	244	238
	61	215	211
	61	325	234
	60	239	205
Every fifteenth day	25	205	176
	25	325	207
	25	239	191
	25	219	196
	25	234	165
	24	201	198
	24	215	211
	24	195	183
	24	188	173
	24	195	169
	24	160	154
	24	244	199
	24	215	201
	24	179	171
	24	238	205

that differ from site to site and year to year. Table 29 indicates the
90th percentile for the sampling schedules used in Table 28.

Indicators of Spread

In addition to an indication of typical and peak values, it is also desirable to have a measure of variability. The customary statistics for this
purpose are the arithmetic standard deviation and the geometric standard
deviation. Range of percentiles can also be used depending upon the desired use of the data summary. The basic formulas for the arithmetic
and the geometric standard deviations are given below.

Let X_1, X_2, ..., X_n be a set of n observations.

Then the arithmetic standard deviations is:

$$s = \frac{1}{n} \sum_{i=1}^{n} (x_i - \bar{x})^2 \Big.^{1/2} \quad \text{where} \quad \bar{x} = \frac{1}{n} \sum_{i=1}^{n} X_i$$

Table 29. Geometric Means, Medians and 90th Percentile Values for Sampling Data

Sampling Schedule	Observations	Geometric Mean	Median	90th Percentile
Every day	365	102.6	97	171
Every sixth day	61	99.8	105	162
	61	95.2	93	155
	61	113.6	113	188
	61	107.2	101	177
	61	106.4	105	171
	60	94.7	94	158
Every fifteenth day	25	100.2	111	175
	25	114.6	121	178
	25	125.0	130	189
	25	104.9	96	192
	25	100.8	105	148
	24	99.8	90	190
	24	104.4	98	177
	24	102.4	99	171
	24	92.1	95	143
	24	100.8	96	162
	24	92.0	88	140
	24	104.6	97	186
	24	107.2	109	173
	24	94.1	94	162
	24	99.6	98	165

and the geometric standard deviation is:

$$s_g = EXP \left[\frac{1}{n} \sum_{i=1}^{n} (\ln X_i - \ln M_g)^2 \right]^{1/2}$$

where M_g is the geometric mean.[1]

FREQUENCY ANALYSIS

A number of investigators,[2-6] primarily Larsen and Zimmer, have shown that short-time averaged air pollution concentrations (from 5-minute to 24-hour averages) sampled over a much longer period of time (*i.e.*, weeks, months or years) tend to be lognormally distributed within the longer sampling interval. This finding leads to statistical applications which allow air sampling data to be analyzed to determine the maximum pollution concentration that probably occurred during a survey, even though the maximum value itself was not observed. This is especially important in light of the fact that most sampling schedules are designed to collect data intermittently (*i.e.*, every other day, every third day or randomly selected days, etc.). The actual days that maximum concentrations occur cannot be determined, but the frequency of occurrence of high concentrations can.

Many different types of air pollutants have been observed to approximate lognormal distributions. Described simply, this means that most pollutants frequently exhibit many low concentrations, a significant number of moderate concentrations and a relatively few number of extreme, peak or maximum concentrations. The histogram of Figure 130 illustrates the relative frequency of occurrence of one-hour average carbon monoxide concentrations observed at the Washington, D.C., CAMP station from 1962 to 1968.[2] The shape of the histogram, which is called "skewed to the right," is typical for lognormally distributed data.

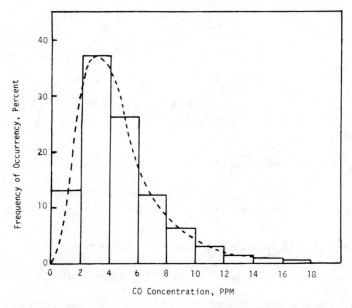

Figure 130. Histogram of 1-hr average CO concentrations, Washington, D.C. (CAMP).

A more illustrative method of plotting the data is shown in Figure 131. This curve illustrates the cumulative frequency distribution of the data when plotted on logarithmic versus probability (log-probability) graph paper. When plotting the cumulative frequency distribution of the data, each point represents the cumulative frequency of equaling or exceeding the specific pollutant concentration. Plotting the cumulative values on log-probability graph paper is, in effect, a subjective method of testing whether the data is indeed lognormally distributed. If the points

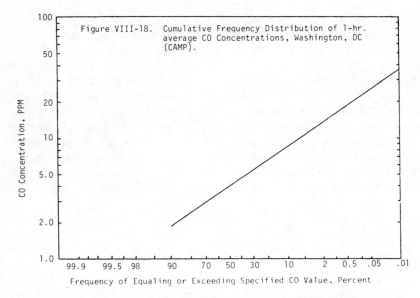

Figure VIII-18. Cumulative Frequency Distribution of 1-hr.
average CO Concentrations, Washington, DC
(CAMP).

Frequency of Equaling or Exceeding Specified CO Value, Percent

Figure 131. Cumulative frequency distribution of 1-hr average CO concentrations, Washington, D.C. (CAMP).

approximate a straight line, then the data is lognormally distributed. If the locus of points form a curved line, the data deviates from lognormality. The types of distributions likely to be encountered have been described by Beaton *et al.*[7] and are illustrated in Figure 132.

COMPARING AIR QUALITY DATA TO STANDARDS

Using Lognormal Cumulative Frequency Distribution Plots

Once the data has been plotted on log-probability graph paper, the frequency of equaling or exceeding any specified concentration can be determined. In Figure 131 the carbon monoxide concentration equals or exceeds 10 ppm 7% of the time, and 20 ppm 0.05% of the time. This is useful when comparing the air quality data to the NAAQS for carbon monoxide. The NAAQS for a 1-hr average concentration of carbon monoxide is 35 ppm, not to be exceeded more than once per year. The allowable frequency of exceeding the standard can be expressed as a percentage by dividing one hour by the number of hours in a year (*i.e.,* 8760):

$$\frac{1}{8760} = 0.00685\%$$

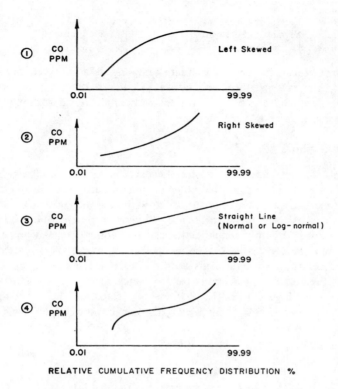

① CO PPM   Left Skewed

0.01   99.99

② CO PPM   Right Skewed

0.01   99.99

③ CO PPM   Straight Line (Normal or Log- normal)

0.01   99.99

④ CO PPM

0.01   99.99

RELATIVE CUMULATIVE FREQUENCY DISTRIBUTION %

Curves 1 and 2 probably indicate lack of data measurements.
Curve 3 is a normal or log-normal distribution of random data.
Curve 4 probably indicates data taken from different populations.

Figure 132. Data plotted on probability or log-probability paper.

To check for compliance at the Washington station, the frequency that 35 ppm is equaled or exceeded can be read in Figure 131. Since this frequency equals 0.01%, which is greater than 0.00685%, this station exceeds the 1-hr NAAQS for carbon monoxide. When using the procedure, care must be taken that air sampling data is plotted at the correct frequency: the data must first be ranked from the highest observed concentration to the lowest, and then the empirical equation, developed by Larsen,[2] used to determine the correct plotting position of each data point.

$$f = 100\% \ \frac{r - 0.4}{1}$$

where: f = the plotting frequency, in %
        r = the rank order (highest, second, third, fourth, etc.)
        n = the total number of samples.

This technique can be used to evaluate sampling data collected over a period of time as short as several weeks. However, the maximum concentration predicted applies only for the sampling interval within which the data was collected.

Short-Cut Methods

It is not always necessary to plot the data on log-probability graph paper in order to determine the expected maximum concentration. A short-cut method of calculating the value without plotting can be applied if the data is believed to be lognormal. Lognormally distributed data can be described by two simple statistics—the geometric mean and the standard geometric deviation. The geometric mean and the standard geometric deviation are equivalent to the intercept and the slope, respectively, of the straight-line cumulative frequency distribution of the data as illustrated in Figure 132. Using these two statistics, the concentration equaled or exceeded at any frequency, C_f, can be determined using the following equation:

$$C_f = M_g \, S_g^{\,Z_f}$$

where: M_g = the geometric mean of the data
        S_g = the standard geometric deviation of the data
        Z_f = the number of deviations from the mean equivalent to the frequency, f

Table 30 lists the appropriate values for Z_f for frequencies ranging from 0.001% to 99.999%.

Saltzman[7] has prepared a nomograph (see Figure 133) which can be used instead of the above calculation, to determine the concentration exceeded at various frequencies. Given the standard geometric deviation of the data, Figure 133 can be used to determine the frequency at which the geometric mean concentration times Saltzman's "deviation factor" is exceeded. If frequencies equal to the NAAQS are used, then the results can be compared to the standards. Figure 133 is applicable only if the distribution of values follows the normal or lognormal distribution precisely, and if the sampling is random.

If the data have already been plotted on log-probability graph paper, a method can be used to determine the geometric mean and the standard geometric deviation that makes use of the graph. First, the geometric mean or median value can be taken from the plot of the data, since it

Table 30. Plotting Position of Extreme Concentrations and Percentiles for
Selected Averaging Times

Averaging Time		No. of Samples in Year	Plotting Position	
	hr		Frequency (60%/N), Percentage of Time	No. of Standard Deviations (z) from Median
1 sec	0.000278	31,500,000	0.0000019	5.50
1 min	0.0166	525,000	0.0001142	4.73
5 min	0.0833	105,000	0.000571	4.39
8.8 min	0.146	60,000	0.001	4.27
10 min	0.166	52,500	0.001142	4.24
15 min	0.25	35,000	0.001715	4.14
30 min	0.5	17,500	0.00343	3.98
1 hr	1	8,760	0.00685	3.81
1.46 hr	1.46	6,000	0.01	3.72
2 hr	2	4,380	0.0137	3.63
3 hr	3	2,920	0.02055	3.53
8 hr	8	1,095	0.0548	3.26
12 hr	12	730	0.0822	3.14
14.6 hr	14.6	600	0.1	3.09
1 day	24	365	0.1644	2.94
2 day	48	183	0.328	2.72
4 day	96	91	0.657	2.48
5.9 day	146	60	1	2.33
7 day	168	52	1.153	2.27
14 day	346	26	2.31	1.99
1 mo	730	12	5	1.64
2 mo	1460	6	10	1.28
3 mo	2190	4	15	1.04
6 mo	4380	2	30	0.52
1 yr	8760	1	50	0.00

is equal to the point where the line crosses the 50th percentile. The
geometric standard deviation, equal to the slope of the line, can be determined by dividing the 84th percentile value (located one standard
deviation from the median) by the 50th percentile value. The geometric
mean and the standard geometric deviation can then be used to determine
extreme concentration values.

AVERAGING TIME ANALYSIS

Air quality data is often collected using sampling intervals of one hour,
resulting in a large quantity of data of one-hour average concentrations
but little data having longer averaging time periods. Larsen[2] has developed

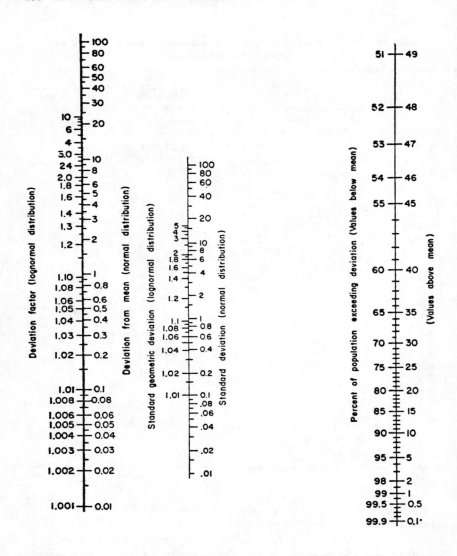

Figure 133. Cumulative percent of population exceeding concentrations having various deviations from the mean. For a normal distribution, use numbers on right sides of center and left scales. A straight line connects related values of the deviation above and below the mean, the standard deviation and the percentage of the population exceeding the value. If desired, the numbers on both the center and left scales may be multiplied by the same factor (*e.g.,* 0.1, 0.01, 0.001, etc.). For a lognormal distribution, use numbers on left sides of center and left scales. A straight line connects related values of the deviation factor (with which to multiply or divide the geometric mean), the standard geometric deviation and the percentage of the population exceeding the value.

a mathematical model which can be used to determine maximum 8-hr or 24-hr concentrations when only 1-hr average data is available. The basic assumptions of the model are:

1. Concentrations are lognormally distributed for all averaging times.
2. The median concentration is proportional to averaging time raised to an exponent (and thus plots as a straight line on logarithmic paper).
3. The arithmetic mean concentration is the same for all averaging times.
4. For the longest averaging time calculated (usually one year), the arithmetic mean, geometric mean, maximum concentration, and minimum concentration are all equal (and thus plot at a single point on Figure 134.
5. Maximum concentration is approximately inversely proportional to averaging time raised to an exponent for averaging time or less than one month.

Data fitting this model can be graphed as shown in Figure 134, which includes the maximum, minimum and frequency of equaling or exceeding specified pollutant concentrations for averaging times ranging from one second to one year. The annual maximum line can be used to determine the maximum concentration for averaging times different from that of the available data. If the standard geometric deviation and the expected annual maximum concentration for a particular averaging time are known, the expected annual maxima for other averaging times can be estimated[8] by using the equation:

$$C_{max} = (C_{MAX - HR}) \, t^q$$

where: C_{max} = the expected annual maximum concentration for a particular averaging time desired

C_{MAX-HR} = the expected maximum 1-hr average concentration

t = the particular averaging time desired in hours

q = the coefficient for transforming averaging time; q is a function of the SGD (Standard Geometric Deviation) of the data and equal to the slope of the line on log-log paper

Values for q are listed in Table 31 for 1-hr average SGD values ranging from 1.00 to 4.99.

To illustrate the equation, assume that the maximum 1-hr average and the standard geometric deviation of carbon monoxide concentrations have been determined for a particular size, equal to 20 ppm and 2.65, respectively. The maximum 8-hr average concentration can be estimated as follows:

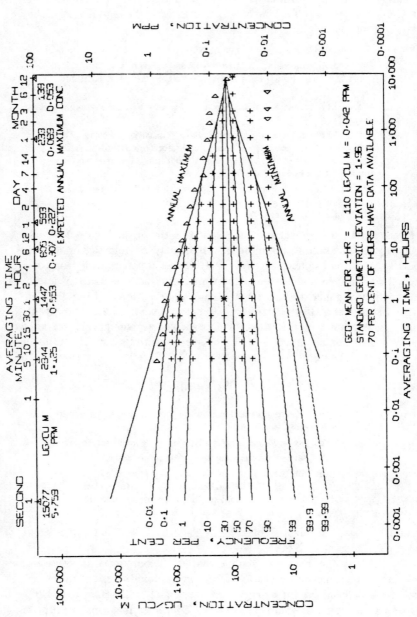

Figure 134. Computer plot of concentration versus averaging time and frequency for sulfur dioxide at site 256, Washington, D.C., December 1, 1961 to December 1, 1968.

Table 31. Slope of Maximum Concentration Line for 1-hr Average Standard
Geometric Deviations from 1.0 through 4.99

SGD[a]	0.00	0.01	0.02	0.03	0.04	0.05	0.06	0.07	0.08	0.09
1.00	0.000	-0.001	-0.008	-0.012	-0.017	-0.021	-0.025	-0.029	-0.034	-0.038
1.10	-0.042	-0.046	-0.050	-0.054	-0.057	-0.061	-0.065	-0.069	-0.072	-0.076
1.20	-0.080	-0.083	-0.087	-0.090	-0.094	-0.097	-0.101	-0.104	-0.107	-0.111
1.30	-0.114	-0.117	-0.121	-0.124	-0.127	-0.130	-0.133	-0.136	-0.139	-0.142
1.40	-0.145	-0.148	-0.151	-0.154	-0.157	-0.160	-0.163	-0.165	-0.168	-0.171
1.50	-0.174	-0.176	-0.179	-0.182	-0.184	-0.187	-0.190	-0.192	-0.195	-0.197
1.60	-0.200	-0.202	-0.205	-0.207	-0.210	-0.212	-0.214	-0.217	-0.219	-0.222
1.70	-0.224	-0.226	-0.229	-0.231	-0.233	-0.235	-0.238	-0.240	-0.242	-0.244
1.80	-0.246	-0.248	-0.251	-0.253	-0.255	-0.257	-0.259	-0.261	-0.263	-0.265
1.90	-0.267	-0.269	-0.271	-0.273	-0.275	-0.277	-0.279	-0.281	-0.283	-0.285
2.00	-0.287	-0.288	-0.290	-0.292	-0.294	-0.296	-0.298	-0.299	-0.301	-0.303
2.10	-0.305	-0.307	-0.308	-0.310	-0.312	-0.314	-0.315	-0.317	-0.319	-0.320
2.20	-0.322	-0.324	-0.325	-0.327	-0.329	-0.330	-0.332	-0.333	-0.335	-0.337
2.30	-0.338	-0.340	-0.341	-0.343	-0.344	-0.346	-0.347	-0.349	-0.350	-0.352
2.40	-0.353	-0.355	-0.356	-0.358	-0.359	-0.361	-0.362	-0.364	-0.365	-0.366
2.50	-0.268	-0.269	-0.271	-0.372	-0.373	-0.375	-0.376	-0.378	-0.379	-0.380
2.60	-0.381	-0.383	-0.384	-0.386	-0.387	-0.388	-0.390	-0.391	-0.392	-0.393
2.70	-0.395	-0.396	-0.397	-0.398	-0.400	-0.401	-0.402	-0.403	-0.405	-0.406
2.80	-0.407	-0.408	-0.410	-0.422	-0.412	-0.413	-0.414	-0.415	-0.417	-0.418
2.90	-0.419	-0.420	-0.421	-0.422	-0.424	-0.425	-0.426	-0.427	-0.428	-0.429
3.00	-0.430	-0.431	-0.432	-0.434	-0.435	-0.436	-0.437	-0.438	-0.439	-0.440
3.10	-0.441	-0.442	-0.443	-0.444	-0.445	-0.446	-0.447	-0.448	-0.449	-0.450
3.20	-0.451	-0.452	-0.453	-0.454	-0.455	-0.456	-0.457	-0.458	-0.459	-0.460
3.30	-0.461	-0.462	-0.463	-0.464	-0.465	-0.466	-0.467	-0.468	-0.469	-0.470
3.40	-0.471	-0.472	-0.473	-0.474	-0.475	-0.476	-0.477	-0.477	-0.478	-0.479
3.50	-0.480	-0.482	-0.482	-0.483	-0.484	-0.485	-0.485	-0.486	-0.487	-0.488
3.60	-0.489	-0.490	-0.491	-0.492	-0.492	-0.493	-0.494	-0.495	-0.496	-0.497
3.70	-0.497	-0.498	-0.499	-0.500	-0.501	-0.502	-0.502	-0.503	-0.504	-0.505
3.80	-0.506	-0.506	-0.507	-0.508	-0.509	-0.510	-0.510	-0.511	-0.512	-0.513
3.90	-0.514	-0.514	-0.515	-0.516	-0.517	-0.517	-0.518	-0.519	-0.520	-0.520
4.00	-0.521	-0.522	-0.523	-0.523	-0.524	-0.525	-0.526	-0.526	-0.527	-0.528
4.10	-0.529	-0.529	-0.530	-0.531	-0.531	-0.532	-0.533	-0.533	-0.534	-0.535
4.20	-0.536	-0.536	-0.537	-0.538	-0.539	-0.539	-0.540	-0.541	-0.542	-0.542
4.30	-0.543	-0.543	-0.544	-0.545	-0.545	-0.546	-0.547	-0.547	-0.548	-0.549
4.40	-0.549	-0.550	-0.551	-0.551	-0.552	-0.553	-0.553	-0.554	-0.555	-0.555
4.50	-0.556	-0.556	-0.557	-0.558	-0.558	-0.559	-0.560	-0.560	-0.561	-0.562
4.60	-0.562	-0.563	-0.563	-0.564	-0.565	-0.565	-0.566	-0.566	-0.567	-0.568
4.70	-0.568	-0.569	-0.569	-0.570	-0.571	-0.571	-0.572	-0.572	-0.573	-0.574
4.80	-0.574	-0.575	-0.575	-0.576	-0.576	-0.577	-0.578	-0.578	-0.579	-0.579
4.90	-0.580	-0.580	-0.581	-0.582	-0.582	-0.583	-0.583	-0.584	-0.584	-0.585

[a]Standard geometric deviation for a particular slope is the sum of the left and
top margin values.

$$C_{max\ 8\text{-hr}} = C_{max\ 1\text{-hr}}\ (8\ hr)^{-0.388}$$
$$= 20\ ppm\ (8)^{-0.388}$$
$$= 8.9\ ppm$$

VALIDATING MATHEMATICAL SIMULATION MODELS

Analytical Methods

Methods for validating mathematical models may employ either an analogical approach or a statistical approach. In an analogical approach, similarities and differences between field measurements and mathematical predictions are investigated using arithmetic, graphic, or other methods. A statistical approach means that the similarities and differences between variables are quantified and described statistically. Analogical approaches to model validation are straightforward and provide insight into the nature of the model and the field situations being studied. One disadvantage of this approach is that it can be awkward when analyzing large quantities of data. In some cases, certain relationships identified using analogical methods can be tested using statistical methods in order to quantify the significance of the observed relationship.

Analogical Approach

Basically, the analogical approach consists of looking for similarities of relationships between variables. In the case of model validation, the most logical first step is to compare the overall model results with the pollutant concentrations observed in the field. One of the ways of evaluating the relationship is to plot the measured pollutant concentration versus the concentration predicted by the model (see "scattergram" plot in Figure 135). The fact that the data deviates considerably from the desired straight-line relationship indicates that the field measurements do not readily verify the overall accuracy or precision of the model.[9]

If the overall accuracy of the model is questionable, the data can be separated into a number of data sets which can be used to test the model under more specific conditions. For example, when the data plotted in Figure 135 is separated according to wind direction (*i.e.,* winds parallel to the roadway and winds perpendicular to the roadway) two new scattergrams can be prepared (see Figures 136 and 137). These two new curves indicate that the relationship between the model predictions and the measured values are consistently different for parallel wind versus perpendicular wind conditions. If this model were to be used to

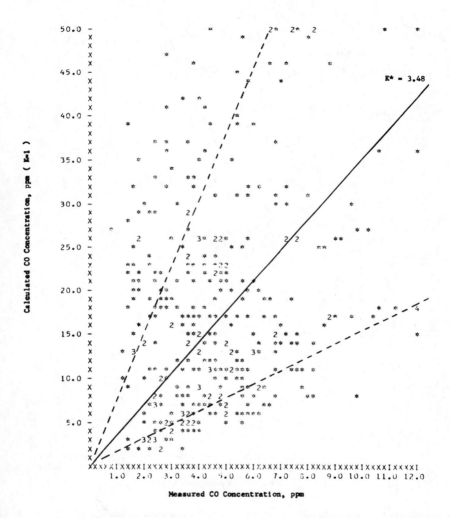

Figure 135. Calculated CO versus measured CO for all data.

assess the air quality impact of a proposed transportation facility, then it might be desirable to adjust the model output values depending on wind direction.

How well the model predicts the decrease in concentrations at greater distances from the highway can also be determined using graphic methods. Figures 138 and 139 illustrate how nondimensionalized pollutant concentrations can be plotted against distance from the road edge.

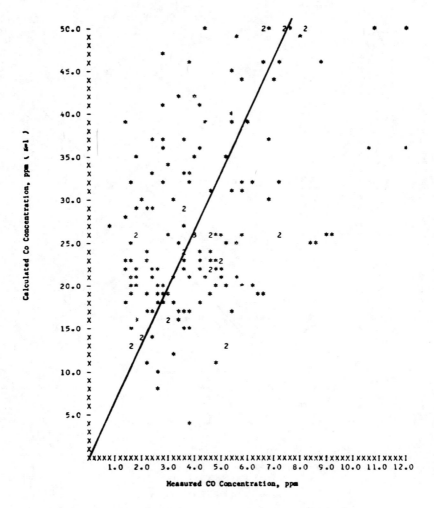

Figure 136. C_c versus C_m for all crosswind data.

Note that the average of the measured values falls within the values predicted by the model for crosswind conditions, but outside the predicted
values under parallel wind conditions. The graphical techniques shown
here are only examples of the many ways various components of a
mathematical simulation model can be checked using plots of field
data.

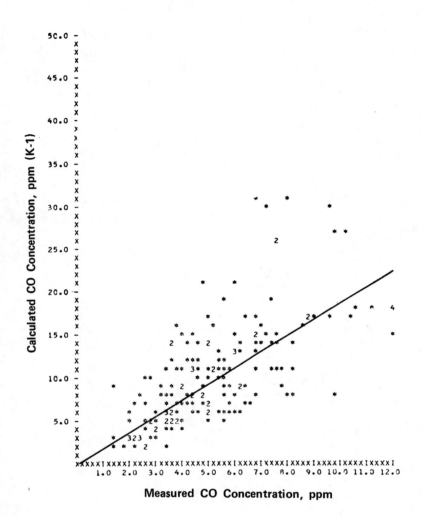

Figure 137. C_c versus C_m for all parallel wind data.

Statistical Approach—Regression Analysis

How closely a mathematical simulation model fits the observed results of a field investigation can be determined using statistical techniques. Probably the most appropriate for model validation studies is regression analysis. Regression statistics are used to estimate the mathematical relationship of one variable with another by expressing the one in terms of

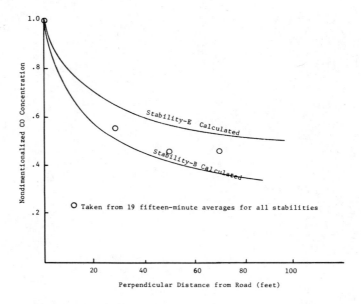

Figure 138. Horizontal profile of nondimensionalized CO concentrations referenced to mixing cell for crosswinds (157, 202, 225).

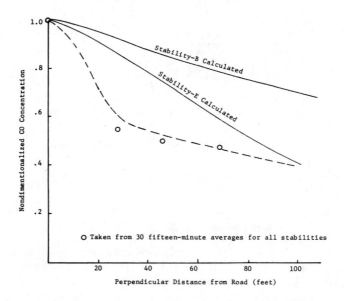

Figure 139. Horizontal profile of nondimensionalized CO concentrations referenced to mixing cell for parallel winds (135 and 270).

a linear (or a more complex) function of the other. In the case of model validation, one variable might be the measured pollutant concentration while the other might be the predicted concentration. Regression techniques could be used to describe the mathematical relationship between the measured value and the predicted value, as well as the variability or amount of error observed for the relationship. An example of the use of regression statistics to validate a highway line source model has been presented by Habegger et al.[10] Figure 140 illustrates the linear relationship between the observed CO concentrations determined for a section of Interstate 55 in Chicago, Illinois. The equation of the straight line was determined using simple linear least square regression techniques and is of the form:

$$Y_i = \alpha + \beta X_i$$

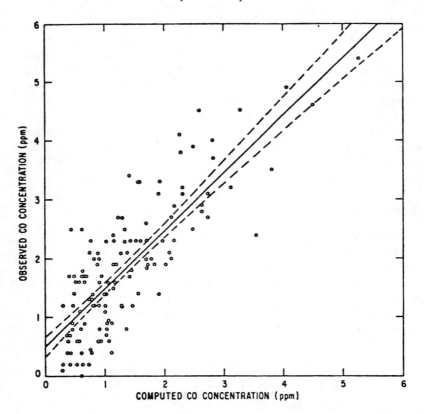

Figure 140. Linear regression line (solid line) with 90% confidence bounds (dashed line) for observed and computed CO concentrations near the at-grade roadway.

where Y_i is the ith measured value, and X_i is the corresponding i^{th} computed value. If the model and measured values agree completely, then the Y-axis intercept, α, equals zero and the slope of the regression line, β, equals unity. Also shown in Figure 140 is the 90% confidence band for the line. Assuming a normal distribution for the errors, for each computer value, X, the mean value for Y is said to lie within this band with 90% confidence.

In addition to determining the equation of the confidence limits that best fit the data, it is also useful to calculate a statistic that describes how much of the variability in the data is "removed" or explained by the regression line. This statistic is the correlation coefficient, r, which is a measure of how well the regression line fits the data. Calculated values for the correlation coefficient, r, range from zero to a maximum value of unit. An r = 1 means that all experimental values lie on the regression line.

Using calculated values for α, β, and r, the validity of a mathematical model can be tested under various conditions, or several different models can be compared. The "best" model performance would be indicated by α values approaching zero, with r and β values approaching unit. For example, Table 32 illustrates 17 different model comparisons presented by Habegger *et al.*[10] From an inspection of α, β and r values, Model 1 was proposed as the best. A disaggregate analysis was then performed on Model 1 by partitioning the data according to wind speed and direction, and calculating the regression statistics for each data set.   Table 33 summarizes the results of the disaggregate analysis performed.

Significance Testing

Regression analysis can be performed on any matched-pairs type of data set; however, the significance of the regression will depend on the number of data points available, and on how well all the data fit a linear relationship. The significance of a regression can be tested by comparing the ratio of the mean square of the y variations due to the regression, divided by the mean square of the y deviations from the regression line, the F-statistic.

Equations Used in Regression Analysis

The equations for regression analysis are:

Slope of the regression line (regression coefficient), β:

$$\beta = \frac{\sum\limits_{i}^{n} (X_i - \bar{X})(Y_i - \bar{Y})}{\sum\limits_{i}^{n} (X_i - \bar{X})^2}$$

Table 32. Experimental Validation of Initial Dispersion Models for Steady-State Gaussian Plumes in the Vicinity of an At-Grade Roadway

Model No.	No. Lines in Model	Initial Source Dimensions			Linear Regression Analysis[b]				Model Average[e] (ppm)
		Location[a]	Height (ft)	Width (ft)	α(ppm)	β	R²[c]	ΔR Significance[d]	
1	1	Edge	20	20	0.490	0.994	0.6526	—	1.34
2	1	Edge	20	0	0.495	0.987	0.6529	33.2%	1.34
3	1	Edge	0	0	0.659	0.611	0.693	77.8%	1.91
4	1	Edge	10	0	0.568	0.780	0.684	95.0%	1.61
5	1	Edge + 10'	10	0	0.596	0.713	0.685	93.6%	1.72
6[f]	1	Edge	10	0	0.528	0.573	0.596	100.0%	2.26
7	1	Center	20	139.5	0.442	1.35	0.619	93.6%	1.02
8	2	Edge	10	0	0.544	1.01	0.645	26.6%	1.27
9	2	Edge	10	0	0.483	1.22	0.624	88.4%	1.10
10	2	Edge	20	20	0.479	1.22	0.623	89.0%	1.10
11	2	Edge + 10'	10	0	0.566	0.939	0.647	19.8%	1.34
12[f]	2	Center	10	40	0.514	1.12	0.635	52.0%	1.16
13[f]	2	Center	10	40	0.554	0.725	0.537	100.0%	1.75
14	2	Center	0	40	0.576	0.953	0.645	19.8%	1.31
15	2	Center	10	0	0.518	1.13	0.636	51.0%	1.16
16	2	Center	20	40	0.463	1.33	0.616	93.2%	1.02
17	2	Center	10	50	0.513	1.13	0.636	52.0%	1.16

[a]With a single line source, center refers to placement of the line source at the center of the median strip, and edge refers to placement at the downwind edge of the total roadway. For two lines, center refers to placement at the centerlines of the inbound and outbound traffic lanes, and edge is placement at the downwind edge of each of the two lanes. Edge + 10' indicates placement of the line sources 10 ft downwind of the dge.

[b]Results of least-squares fit to model $Y_i = \alpha + \beta X_i$ where Y_i, X_i (ppm) are experimental and computed concentrations, respectively.

[c]Correlation coefficient.

[d]Assuming a normal distribution for differences between experimental values and the linear regression line, ΔR significance is the probability that the difference in R^2 between the given model and Model 1 is not due to chance alone.

[e]Average of the computed values, X_i. The average of the experimental values, Y_i, is 1.83 ppm.

[f]A neutral stability class (D) was assumed for all cases in these models.

Table 33. Experimental Validation of Reference Model (Model 1, Table 32)
for Various Classes of Wind Speed and Direction

Wind Speed (mph)	Wind[a] Direction	Linear Regression Analysis[b]			Data Average (ppm)	Model Average (ppm)	No. Data Points
		α (ppm)	β	$R^{2\,c}$			
0-4	0-59	1.13	0.814	0.703	3.24	2.59	15
0-4	60-90	-0.604	1.51	0.880	2.13	1.81	17
4.5-8	0-59	0.580	0.910	0.571	1.65	1.18	60
4.5-8	60-90	-0.099	1.48	0.488	1.44	1.04	18
> 8	0-59	0.777	0.797	0.156	1.44	0.83	21
> 8	60-90	–	–	–	1.40	1.15	2

[a]The angle between the wind direction and roadway.
[b]Results of least-squares fit to model $Y_i = \alpha + \beta X_i$ where Y_i and X_i (ppm) are experimental and computed concentrations, respectively.
[c]Correlation coefficient.

Intercept of the regression line, α:

$$\alpha = \bar{Y} - \beta \bar{X}$$

Correlation coefficient, r:

$$r = \frac{\Sigma\,(\alpha + \beta X_i - \bar{Y})^2}{\Sigma\,(Y_i - \bar{Y})^2}$$

Confidence limits on the line, L_i:

$$L_i = \alpha + \beta X_i \pm t_{a,v} \qquad W_{Y \cdot X}^2\,\frac{1}{n} + \frac{(X_i - \bar{X})^2}{\Sigma\,(X - \bar{X})^2}$$

$$S_{Y \cdot X}^2 = \Sigma\,(\alpha + \beta X_i - Y_i)^2$$

where: X_i = the ith value of X
$\bar{X}$ = the arithmetic mean of X
Y_i = the ith value of Y
n = the number of data points
$t_{a,v}$ = t-statistic from statistical tables
$S_{Y \cdot X}^2$ = the mean square unexplained deviation of Y's from the regression line.

DETERMINING SAMPLE SIZE

Statistical methods for determining the number of samples needed to accurately define the mean and maximum pollution concentrations expected to occur during a year have been presented in the literature by several authors.[11-13] In general, these methods assume that the samples collected are representative of the total population, which is either normally or lognormally distributed, and that each sample is chosen randomly. Under these conditions, the simplified methods presented by Saltzman[13] can be conveniently used to determine the total number of samples needed.

Sample Sizes and Confidence Intervals for the Mean

Saltzman[13] assumes that the population to be sampled is infinite and applies the statistical t_{n-1} distribution to determine the confidence intervals of the mean.

$$\text{Confidence Interval} = \overline{X} \pm t_{n-1} \, S/n$$

where: $\overline{X}$ = mean value of the samples
  S = standard deviation of the samples
  n = number of samples
  t_{n-1} = a tabulated statistic for various confidence coefficients and degrees of freedom

For more than 150 samples, t_{n-1} has a value close to that of the normal distribution statistic, $Z = 1.96$ for 95% confidence interval and 2.576 for 99%). The nomograph illustrated in Figure 141 permits simple application to lognormal as well as normal populations.

Example: The standard geometric deviation of a lognormal distribution is 1.5. How many samples should be collected to establish the geometric mean within ± 5% with 95% confidence? How many for 99% confidence? Answer: Draw a straight line through 1.05 on the left side of the left scale and 1.5 on the left side of the center scale. The intersections on the two right scales are 250 and 440, signifying that if groups of 250 samples are collected, the geometric mean of 95% of the groups would be within the interval of the true geometric mean multiplied or divided by 1.05. Also, if groups of 440 samples are collected, the same agreement would be obtained for 99% of the groups.

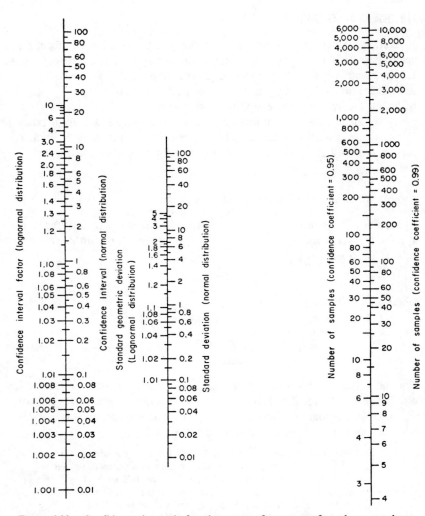

Figure 141.   Confidence intervals for the mean of a group of random samples.
For a normal distribution, use numbers on right sides of center and left scales. A
straight line connects related values of the confidence interval (above and below the
mean), the standard deviation, and the number of samples (n) in the group for either
0.95 or 0.99 confidence coefficients (calculated from the t_{n-1} distributions). If
desired, the numbers on both the standard deviation scale and on the confidence
interval scale may be multiplied by the same factor (*e.g.,* 0.1, 0.01, 0.001, etc.).
For a lognormal distribution, use numbers on left sides of center and left scales.
A straight line connects related values of the confidence interval factor (with which
to multiply or divide the geometric mean), the standard geometric deviation, and
the same scales for number of samples in the group.

Sample Sites and Confidence Intervals for Extremes

Another important problem is the design of a sampling program to determine conformity with specified air quality standard. For a given proportion (p) in the population exceeding a specified concentration, the probability of finding c or fewer samples exceeding the same concentration in a total number n of samples is given by:

$$\text{Probability } (m \leqslant c) = \sum_{m=0}^{c} \frac{n!}{m!(n-m)!} \, p^m (1-p)^{n-m}$$

m = number of positive samples, assumed to have all integral values from 0 to C.

This calculation can be performed using the nomogram presented by Saltzman.

> Example: Design a sampling program so that the probability of acceptance is > 0.95 if actually $< 2\%$ of the population exceeds the critical value, but is < 0.10 if actually $> 8\%$ do. Answer: (See illustration on the upper right of Figure 142). Draw a line connecting 2% on the left scale with 0.95 on the right scale and another one connecting 8% with 9.10. Their intersection in the center area gives values c = 4, n = 98. Thus, 98 samples should be collected and no more than 4 are allowed to exceed the critical concentrations.

It must be noted, however, that this calculation is based upon a random sampling program. If the peak values are known to occur at certain specified times, the procedure may be modified to collect most samples during those periods to provide greater assurance that such peaks are not occurring, or to determine the proportion with greater accuracy. For example, for peak oxidant sampling, the day may be divided into an 8-hr period and a 16-hr period, such as 9:00 a.m. to 5:00 p.m. and 5:00 p.m. to 9:00 a.m. Only a few samples need be collected during the later period. Thus, considerable analytical work can be saved with almost no increase in confidence intervals. The proportions in each interval may be weighted by the length of the interval and combined to give the proportion in the entire day.

Sampling from Finite Populations

Hale[11] has presented methods that assume that the population from which samples are taken is finite. In general, the number of samples required according to Hale's methods are less than for Saltzman's. For a lognormal distribution, the number of samples, n, required to determine

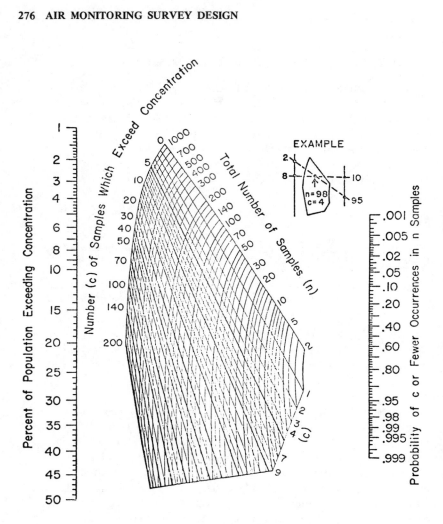

Figure 142. Design of a sampling program for quality control. Related variables are connected with a straight line. For proportions below 1%, divide numbers on left scale by a convenient integer (such as 19), and multiply numbers on n scale by the same integer. For proportions above 50%, the nomogram may be used by regarding the left scale and the c scale as representing the numbers of samples equal to or less than the critical concentration. This chart also may be used to estimate the confidence intervals for proportions in the population by drawing lines through point [(c - 1/2), n]; in this case the right scale represents probability that the true proportion is greater than the value intercepted on the left scale.

tolerance and confidence interval given by Hale's equation is:

$$n = \frac{NZ^2 1n^2 S_g}{N\ 1n^2(P+1) + Z^2 1n^2 S_g}$$

where: N = population size

Z = normal deviate corresponding to the upper percentage point for a specified level of confidence. For a 95% level Z = 1.96.

S_g = standard geometric deviation of samples

P = fraction of the geometric mean by which it can differ from the true geometric mean with specified probability.

Using this equation, Figures 143 and 144 have been prepared. These figures can be used to determine the number of samples needed as a function of the standard geometric deviation and the size of the population being sampled. Figure 143 should be used to determine the geometric mean within ± 10% (at 95% confidence) and Figure 144 should be used for a ± 20% tolerance (at 95% confidence).

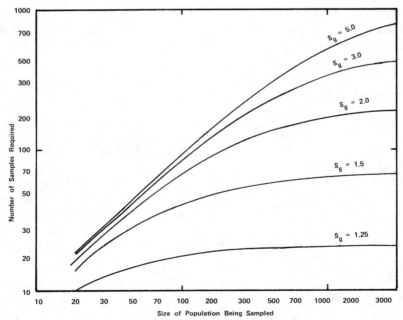

Figure 143. Number of samples required to determine the geometric mean within ± 10% with 95% confidence for various population sizes (Z = 1.96 % P = 0.10).

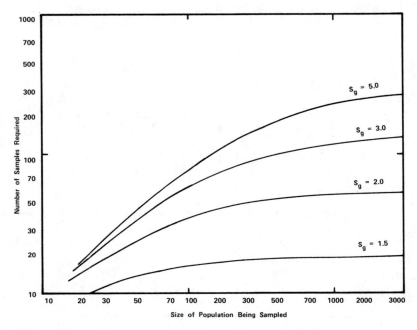

Figure 144. Number of samples required to determine the geometric mean within ± 20% with 95% confidence for various population sizes (Z = 1.96 & P = 0.20).

REFERENCES

1. *Guidelines for Air Quality Maintenance Planning and Analysis, Vol. II: Air Quality Monitoring and Data Analysis*, EPA, Office of Air Quality Planning and Standards, Publication No. EPA-450/ 4-74-012 (September 1974).
2. Larsen, R. I. *A Mathematical Model for Relating Air Quality Measurements to Air Quality Standards*, EPA Publication No. AP-89 (November 1971).
3. Zimmer, C. E. and R. I. Larsen. *J. Air Pollution Control Assoc.* 15:565 (1964).
4. Larsen, R. I., C. E. Zimmer, D. A. Lynn and K. G. Blemel. *J. Air Pollution Control Assoc.* 17:85 (1967).
5. Larsen, R. I. *J. Air Pollution Control Assoc.* 19:24 (1969).
6. Altshuller, A. P., G. C. Ortman, B. E. Saltzman and R. E. Neligan. *J. Air Pollution Control Assoc.* 16:87 (1966).
7. Beaton, J. L., A. C. Ranzieri, E. C. Shirley and J. B. Skog. *Air Quality Manual, Vol. II—Analysis of Ambient Air Quality for Highway Projects* (Sacramento, California: California Division of Highways, April 1972).

8. Miller, I. and J. E. Freund. *Probability and Statistics for Engineers* (Englewood Cliffs, New Jersey: Prentice-Hall, Inc., 1965).
9. Noll, K. E. "Air Quality Report: Interstate I-40 Modification between Stratford and Road and Peter's Creek Parkway, Forsyth Co., North Carolina" (August 1973).
10. Habegger, L. J., T. D. Wolska, J. E. Camaioni, D. A. Kellermeyer and P. A. Dauzvardis. "Dispersion Simulation Techniques for Assessing the Air Pollution Impacts of Ground Transportation Systems," Argonne National Laboratory, Contract W-31-109-Eng.-38, (June 18, 1974).
11. Hale, W. E. "Sample Size Determination for the Lognormal Distribution," *Atmos. Environ.* 6(6):419-422 (1972).
12. Hunt, W. F., Jr. "The Precision Associated with the Sampling Frequency of Lognormally Distributed Air Pollutant Measurements," *J. Air Pollution Control Assoc.* 22(9):687-691 (1972).
13. Saltzman, B. E. "Simplified Methods for Statistical Interpretation of Monitoring Data," *J. Air Pollution Control Assoc.* 22(2):90-95 (1972).

GLOSSARY *

Air Monitoring: The process of sampling and analyzing the concentration of air pollutants in ambient air.

Air Sampling: The process by which air samples or air pollution samples are collected.

Ambient Concentrations: Any air pollution concentration that occurs in the outdoor environment. In contrast to concentrations "in-plant" or "in-stack," in-plant concentrations are considered as industrial hygiene problems; in-stack concentrations refer to air pollution emission rates.

Analytical Methods: Methods of chemically analyzing the quantities of concentration of air pollutants in an air sample.

Anemometer: An instrument for measuring and indicating the speed of the wind.

Audit Gas: A primary standard bottle of calibration gas used primarily to calibrate a secondary standard.

Background (near highways): The concentration which, when added to the contribution of the highway project itself and specific local contributions from other major sources, will give the total pollutant concentration. Background is a macroscale or mesoscale phenomenon. To measure background, the site must be outside the influence of microscale air pollution regimes (*i.e.*, nearby sources).

Bias Removed (wind data): Wind direction measurements reported by people frequently contain more observations of the wind directions N, NE, E, SE, etc. and fewer observations of NNE, ENE, ESE, SSE, etc., than would be expected to occur naturally. This error is of human origin, and can be "removed" using statistical methods.

---

*Some of these definitions are from Elfers, L. A. *Field Operations Guide for Automatic Air Monitoring Equipment*, EPA Publication No. PB-202-249.

Calibrate (models):   To adjust mathematically the pollution concentration predicted by a model so as to match the results obtained from model validation field studies.

Calibrating Solution: A solution, containing a known amount of a substance having an effect equivalent to a pollutant concentration, that is passed through the detection component during the static calibration of an analyzer.

Calibration, Dynamic: A calibration of the complete instrument by means of sampling either a gas of known concentration or an artificial atmosphere containing a pollutant of known concentration.

Calibration Gas: Gas used to calibrate an air monitoring instrument.

Calibration, Static: A performance test of the detection and signal presentation components accomplished by using an artificial stimulus such as standard calibrating solutions, resistors, screens, optical filters, electrical signals, etc., which has an effect equivalent to pollutant concentrations.

Calms Distributed (wind data):   Wind speeds less than 1 m/s are often reported as "calm" conditions.   When winds are calm, wind direction measurements are meaningless.   Hence, when wind roses are prepared, the percentage of the time when winds are calm is distributed among all wind directions on a weighted basis.

Collection Efficiency: The amount of substance absorbed or detected divided by the amount sampled.

Data Reduction: The process by which chart records or digital records from an air monitoring instrument are converted from relative values of electrical output to specific ppm or $\mu g/m^3$ air pollution concentration values.

Detector (sensor, transducer):   A device that detects the presence of an entity of interest and indicates its magnitude as the deviation from a reference and converts these indications into a signal (*i.e.,* photometer, infrared holometer, flame ionization detector, etc.).

Diluent Gas: Any gas mixture used to dilute the air pollution concentration in an air sample;   usually zero air or nitrogen.

Dilution Panel: Device used to accurately control flow rates of air pollution and diluent gases which are then mixed together to produce known concentrations of air pollution mixtures to be used for instrument calibration purposes.

Dry Gas Meter (bellow or diaphragm meter):   A device that measures total volume of a gas passed through it without the use of volatile liquids.

Equivalent Method: A method of sampling and analyzing ambient air for an air pollutant that has been demonstrated as equivalent to a reference method in accordance with Part 53, Title 40 of the Code of Federal Regulations (40-FR7044).

Instrument Inlet: The opening at the instrument through which the sample enters the analyzer, excluding all external sample lines, probes and manifolds.

Macroscale: The macroscale air pollution regime is any air mass exhibiting ground level air pollution concentrations that deviate by less than 20% over linear distances greater than 10,000 m. An example of a macroscale regime is regional background air pollution concentrations, which can be fairly homogeneous over linear distances from tens to hundreds of kilometers. This does not mean, however, that there are not locations within the macroscale regime where ambient air pollution concentrations deviate by more than 20% from regional background levels. These deviations can and do occur when mesoscale and microscale air pollution regimes are superimposed upon the regional background air pollution concentration regime.

Mesoscale: The mesoscale air pollution regime is any air mass exhibiting ground level air pollution concentrations that deviate by less than 20% over linear distances between 100 m and 10,000 m. The mesoscale regime represents a "community-size" air mass exhibiting faily homogeneous ground level air pollution concentrations—for example, local background ambient air pollution concentrations that result, especially within urban areas, from the emission of relatively small quantities of air pollutants from a large number of ground level sources (*i.e.*, automobiles, residential and commercial space heating furnaces, and even numerous small industrial sources). These local background concentrations can vary considerably at different locations within an urban area. Concentration gradients greater than 20% per 100 m indicate the presence of a microscale air pollution regime superimposed upon the mesoscale regime.

Microscale: The microscale air pollution regime is any air mass exhibiting ground level air pollution concentrations that deviate by greater than 20% over linear distances up to 100 m. The microscale air pollution regime represents a relatively small air mass exhibiting large variations in ground level air pollution concentrations. This phenomenon usually occurs very near sources of air pollution when the rate of increasing atmospheric dispersion with downwind distance is very great.

NAAQS: National Ambient Air Quality Standards (See *Federal Register* Vol. 36, No. 228, November 25, 1971).

Orifice: An opening through which a fluid can pass. The orifice so shaped as to provide pressure-flow characteristics which can be translated to flow rate. At fluid flow rates approaching sonic velocities, the orifice provides flow characteristics useful for controlling flow.

Primary Standard: A substance with a known property that can be defined, calculated or measured, and that is readily reproducible. The standard may be traceable to the National Bureau of Standards or other accepted standards organization.

Reference Method: A method of sampling and analyzing ambient air for an air pollutant that is specified in the *Federal Register* as appropriate for comparing measurements to the NAAQS.

Rotameter (variable area flow meter): A flow-measuring device that operates at constant pressure and consists of a weight or float in a tapered tube. As the flow rate increases, the float moves to a region of larger area and seeks a new position of equilibrium in the tapered tube, the position being related to flow rate.

Sample: A representative portion or specimen of an entity presented for inspection.

Sample, Integrated: The sum of a series of small samples or a continuous flow of sample collected over a finite time period (from minutes to hours) so as to create a large average sample. An integrated sample is often stored for a period of time before analysis although losses may occur.

Sampling, Continuous: A process in which samples are collected continuously at a known rate.

Sampling, Random: A process in which grab or integrated samples are collected at random intervals.

Sampling, Sequential: A process in which a series of individual grab or integrated samples are collected one after the other at regular predetermined intervals.

Secondary Standard: A substance having a property that is calibrated against a primary standard to a known accuracy.

Span Gas: A calibration gas having a known concentration of air pollution, usually in a concentration range near the highest ambient values expected to be measured.

Spatial Resolution: The ability to resolve the spatial distribution of air pollution concentrations in an area by modeling or monitoring. The spatial distribution of pollution concentrations is affected by the following:

1. **Distance to the Source:** Receptors located at greater distances from ground level sources will exhibit lower air pollution concentrations than receptors located nearer the source. For example, doubling the distance separating source and receptor can decrease the pollutant level by a factor of from 1.5 to 7.7 for F and A stabilities respectively. (Taken from *Workbook of Atmospheric Dispersion Estimates*, by D. Bruce Turner, AP-26.)

2. **Number of Sources:** Receptors located near numerous pollution sources will generally exhibit higher concentrations than receptors located near only a single source. Numerous ground level sources are frequently designated as an "area source" of pollution emissions. Single sources, especially those emitting from tall stacks, are referred to as "point sources" of pollution emissions.

3. **Emission Height:** Sources emitting pollution from tall stacks generally cause much lower pollution concentrations at receptors than sources emitting pollution near ground level. For example, according to Turner's Workbook, increasing the height of the emission point from 10 m to 20 m will result in a 4.3-fold decrease in the maximum ground level concentration under conditions of D stability. Increasing the height of the emission point to 50 m results in a 27-fold decrease in the maximum ground level concentration.

4. **Emission Strength:** Receptors located near sources having high pollution emission rates will generally exhibit higher concentrations than receptors near sources with lower emission rates.

Specific Response: Response of an air pollution analyzer to a single pollutant species without interference from others.

Stock Solution: A solution containing a substance that has an effect equivalent to pollutant concentrations, and which is standardized against a primary or secondary standard.

Temporal Resolution: The ability to resolve the temporal distribution of air pollution concentrations in a study area by modeling or monitoring. The temporal distribution of pollution concentrations is affected by the following:

1. **Sources** of air pollution seldom emit material into the air at a constant and continuous rate. Even industrial sources have different phases in their processes that produce different emissions. Automobiles generate different quantities of air pollutants for different operating modes and speeds. Streets and highways, viewed as an air pollution source, generate different quantities of emissions depending largely on the traffic volume. As the traffic volume increases, so do pollution emissions and the resulting ground level pollution concentration.

2. **Changing wind direction** can have very pronounced effects on the temporal distribution of air pollutants. Changing wind direction

can cause a receptor, located near a source, to be in a downwind position during one hour, then change to an upwind position (relative to the source) during a different hour. Whenever the receptor is located downwind of the source, relatively high concentrations may result. When the receptor is in an upwind position, a much lower concentration may result. In this manner, changing wind direction can affect the temporal distribution of pollution concentrations at a single point. Whenever the wind direction changes, the spectrum of upwind sources will necessarily change, resulting in a different pollution concentration measured at the receptor.

3. **Changing wind speed** can also have a major effect on the temporal distribution of pollutant concentrations. In general, ground level pollution concentrations downwind of a source are inversely proportional to wind speed. That is, as wind speed increases, ambient pollution level normally decreases. In this respect, wind speeds fluctuating from 1 to 10 m/s can have a 10-fold effect on the ground level air pollution concentration. In inland areas, frequently wind speeds are very low during the early morning and early afternoon. These variations in wind speed will produce similar fluctuations in the measured temporal distribution of ambient pollution levels. In this example, the pollution levels would tend to be highest during the early morning, then decrease later in the day as the wind speed increases. Ambient concentrations may also vary due to the effects of wind speed transporting air pollution from distant sources. Frequently, rural environs exhibit low air pollution concentrations throughout most of the day, then during late afternoon or early evening air pollutants emitted hours earlier within an urban area are transported by the wind into the rural areas. When this occurs, the temporal distribution of air pollutants within the rural area is affected by the transport of air pollution by the wind.

4. **Changing atmospheric stability** can produce fluctuation in the ambient air pollution levels by either increasing or decreasing the amount of horizontal and vertical dispersion of air contaminants that occurs. Even when all other factors are held constant, ambient pollution levels can fluctuate as the conditions governing the horizontal and vertical dispersion of air pollutants change. Stable conditions that are common during nighttime and early morning, usually "break up" by late morning or early afternoon when neutral or unstable conditions predominate. When dispersion conditions improve, ground level pollution concentrations generally decrease.

5. **Topography** can effect the temporal distribution of air pollution by affecting air movements that might not otherwise occur. Examples of this are land and sea breezes, drainage winds and channeling of air movements by valleys or street canyons. Each of these conditions can either enhance or inhibit the dispersion of air pollutants. The effect of topography on the temporal

distribution of air pollutants is not truly an independent effect. Its impact is due primarily to the way topography modifies the natural ventilating effects of meteorology. In many cases it is, however, an important factor.

Validate (models):   To test the accuracy of mathematical models used to predict pollution concentrations by comparing predicted concentrations to measured concentrations under equivalent conditions.

Wet-Gas Meter: A volumetric flow-measuring device that measures the total gas volume by entrapping the gas in inverted cups or vanes under a liquid.   The buoyancy of the gas causes a rotation of the cups or vanes that is proportional to the volume that is indicated on a pre-calibrated meter.

Working Gas: A secondary standard bottle of calibration gas used to calibrate air monitoring instruments.

Working Solution: A solution prepared from a stock solution and used for preparing a range of calibrating solutions.

Zero Gas: A calibration gas containing an air pollution concentration very near zero.   Used for calibrating the baseline or "zero" value for an air monitoring instrument.

INDEX

9-7 ↑50% = 3:15. — 7:45

length study = $\dfrac{n}{f(s)}$